PODS

PODS
Wildflowers and Weeds in Their Final Beauty

GREAT LAKES REGION, NORTHEASTERN
UNITED STATES AND ADJACENT CANADA

A Visual Guide
FROM FLOWER...TO POD...TO DRIED ARRANGEMENT

BY JANE EMBERTSON
PHOTOGRAPHY BY JAY M. CONRADER

Charles Scribner's Sons
New York

ACKNOWLEDGMENTS

Many wonderful people have come into my life during the course of writing this book. I would like to express my sincere appreciation to the following people, professionals in their fields, for their generous time and answers to botanical questions:

To Steven Ballou, Department of Botany, University of Wisconsin, Milwaukee, Wisconsin; Harold W. Rock, director of the Wehr Nature Center, Hales Corners, Wisconsin; Roy Lukes, resident naturalist at the Ridges Sanctuary, Bailey's Harbor, Wisconsin; Victoria Nuzzo, botanist, Dane County Highway Department, Madison, Wisconsin; Rosemary Fleming, Dane County naturalist, University of Wisconsin Arboretum, Madison, Wisconsin; Jerry Schwarzmeier, head naturalist, and James H. Riemer, the Retzer Nature Center, Waukesha, Wisconsin.

To Jay M. Conrader, my photographer, for his patience and expertise, and to his wife, Constance Conrader, for her painstaking care of details and her library assistance.

To Ruth M. Weber and Colleen J. Weiler, for their creative dried arrangements.

To Kitty Kohout, LaGrange Park, Illinois, for the use of the Figwort and Sleepy Catchfly flower transparencies; and to the Wehr Nature Center for Western Ironweed and Lance-leaved Goldenrod.

To Ilene Birge, whose editing started me off in the right direction, and to Mabel Solverson, who typed, typed, and retyped.

And most of all to my dear, enduring husband and family who let me take over the entire house, recreation room, tool shed, and garage with hanging bunches of drying pods and 150 arrangements waiting to be photographed.

The work is done. I am grateful to all these remarkable people.

Jane Embertson

1

CONTENTS

For a companion
on my walking trip . . . perhaps
a little butterfly

Shiki

PREFACE

The purpose of this book is to focus on the fall and winter shapes of nature's seed containers, the **pods —wildflowers and weeds in their final beauty.** It is a reference book for those who find beauty in nature's wild growth and wish to *identify* and *use* common wildflower pods and weed pods, adding a fourth season to outdoor study and collecting.

It is a guidebook in full color for visual identification of pods for the outdoor enthusiast—the hiker, camper, cross-country skier, ecology student, country dweller, wildflower gardener, artist, craftsman, and photographer—as well as a handy reference work for garden clubs.

This selection of wild plants is intended to stimulate people to observe the beauty of the pods, to learn where to find them, when to pick them, and how to use them creatively. Pods, dried flowers, and grasses are the ideal answer for winter bouquets when fresh flowers are no longer available.

The visual section is arranged so that early in the season the species can be identified by its *flower* in the *small photo on each page;* its future stage, the *pod,* is in the *large adjacent photo.* This makes it possible to plan ahead, knowing the pod shape to expect from a certain flower. Later in the season, the reverse is true. The pod can be identified by its photo and easily referred to the adjacent photo of the flower. Becoming familiar with the *flower* and the shape of its *pod* is the purpose of this book.

Seedpods that lack decorative or aesthetic form have been omitted. Pods that are rare or protected species have also been omitted in the interest of preservation.

When beauty achieves great subtlety, the Japanese call the effect *shibui,* meaning "restrained elegance." It is hoped that this book will point out the restrained beauty of these wild pods.

AUTHOR'S NOTE

When I moved out to the Wisconsin countryside, I took wildflower courses, went on field trips into marshes and along streams, tramped through woods, open fields, and prairies, and became fascinated by the wild beauty of nature and the continual change of color and pattern in each season. It was in late fall that I became particularly aware of a new beauty, the *seedpods.* With their contrasting shapes of fuzzy heads, hard shiny pods, berries, sprays, spikes, and spears, ranging in hue from pale beige and ocher to deep reddish-brown mahogany, they were a new beauty to behold.

I noted that many of the plants changed shape so drastically from flower stage to pod that they became difficult to identify. For example, species of the Milkweed family have a variety of pods. The most familiar, the Common Milkweed, with its boat-shaped pod on a rugged stalk, has a sturdy, warty gray exterior with a silky, shiny yellow interior. The Whorled Milkweed, by contrast, has a delicate, parchmentlike pod. The open, empty pods resemble butterflies perched on a fragile stem, whereas the Swamp Milkweed has a completely different stem formation.

As a watercolor artist, I found that the beautiful shapes of the seedpods silhouetted against the intense blue of the autumn sky, the warm, earth-toned masses of pods swaying in the fields, and later on the sharp, dark shapes and shadows in the snow became subjects to study. I discovered, however, that I could not relate many of the pods to their original flowers. Local libraries and bookshops could not provide books on identification of seedpods, so my research was launched.

I began with careful observation, returning to areas at different times of the year, identifying the flowers in the spring and summer, noting their habits, subsequently picking their pods in the fall, and confirming original identification. It became a three-year project.

Weeds and wildflowers, like all growing things, have their own characteristics, and it became important to take the time to note which plants were "worth knowing." On one visit to an area there might be little to observe; but on another visit I would be pleasantly surprised to find the place ablaze with new varieties in bloom and pod. The more pods I collected, the more apparent it became that there *was* a special time to pick each species in order to catch them at their peak color and at a time when they would preserve well.

To me, the empty seed container—the pod—is another of nature's works of art, as beautiful as the flower and as unique in its own form.

My increasing interest in wild plants led to further study of their lore. I found that many kinds of weeds were sources of healing potions, salves, drugs, and medicine, as well as ingredients for scents, sachets, dyes, cosmetics, wines, teas, seasonings, and cooking. There are many books available that can be read on a cozy winter evening to pursue an interest in these uses.

The spiritual lift and physical exercise derived from going out to gather, identify, sketch, or photograph the pods in the late fall or winter snow were a delightful bonus. The wildflowers became my outdoor companions; the thrill of discovery and identification became something I wanted to share.

I hope this book will persuade you to share in similar pleasures, to extend your world to the out-of-doors, and to enjoy the wildflower in its final statement.

Savor peaceful moments while pod hunting—it is a gentle experience.

HOW TO USE THIS BOOK

The wildflowers and weeds selected for this guidebook are found throughout the Great Lakes region, the northeastern United States, and adjacent Canada. The common and conspicuous species will be found all through the area where frost or freeze occurs, and some are quite common to the mid-South. Wildflowers and weeds that lack interesting and decorative pods were omitted.

Weather from year to year affects the abundance of specimens and the quality of the pods. A cool spring, a summer with too much or too little rain, strong autumn winds, or a too-early frost—all have an influence on the development of the pods. One year a certain species may be beautiful and abundant, but the next year it may be scarce or not very showy. Also, wildflowers may be annual, perennial, or biennial, each season bringing a new look to the same area.

People with little botanical knowledge will be able to identify the weeds and wildflowers presented in this book by studying the full-color photo of the flower in bloom, noting its color, leaf shape, and the position of leaves on the stem. The pod is photographed without a background to simplify its identification; the pod shape and stem structure should be carefully noted. The third photo on each page illustrates how the pod can be used in a dried floral arrangement. *Each photograph is accompanied by important data for easy reference.*

The text is designed for the layman. The botanical name of the plant is given along with the common name, should further identification or research be desired. (*New Britton and Brown Illustrated Flora of the Northeastern United States and Adjacent Canada,* by H. A. Gleason; *Wild Flowers,* by Homer D. House; and especially easy to use, *Field Guide to Wildflowers,* by Roger Tory Peterson and Margaret McKenny are excellent technical sources for study. *Gray's Manual of Botany,* by M. L. Fernald, is a more highly technical source.) Many wildflowers have several common names depending on locality; therefore, the botanical name provides positive identification. In-depth discussion of the flowers has been left to other guides. The focus of this book is the decorative seed containers.

Flowers are arranged in each section according to the time of the year in which they begin to pod, early or midsummer, early or late fall. It follows that the later the bloom, generally the later in the season the pod will appear. Early blooms have pods ready to pick by early summer, but the late blooms may not be dry enough for picking until mid-October or later. Interesting to note is that generally the early-blooming wildflowers are short. Plants do not need to grow very high in spring to get sunlight. In summer, as the competing plants grow, so do the flowers. By fall, most of the floral species are very tall. Pod picking is a continuous affair extending even into winter when the pods show above the snow.

The photographs of flowers and pods are divided into five sections for easier identification.

Part 1. SUN—Weeds and wildflowers found in open fields, prairies, farmlands, roadsides, disturbed land and vacant lots.

Part 2. WOODS, PARTIAL SHADE—Species that need some shade and more moist conditions.

Part 3. WETLANDS—Flowers, as well as sedges, found in marshes, bogs, lowlands, and shorelines.

Part 4. GRASSES, SEDGES, AND RUSHES—A special section for quick identification of varieties of grasses and a few sedges and rushes that are interesting in shape and texture to add to arrangements. Most are found in open fields. The sedges and rushes like wet areas.

Part 5. WINTER SKELETONS—Pods strong enough to withstand the winter winds and snow.

(The flowers and pods are classified according to where they are *most likely* to be found. Of course there are exceptions. A species found in field and sun may also be found at the edge of the woods.)

DISPELLING MYTHS ABOUT "WEEDS"

Almost everyone thinks kindly of wildflowers. Weeds are more likely to be objects of scorn. To remove the stigma of "just weeds," some myths should be dispelled. Many of the plants that are now called "weeds" were brought here from Europe by the early settlers for herbal uses. They were cultivated, and many still are; but those that were best able to survive with the least care and that readily multiplied *where not wanted* became known as "weeds." Unfortunately, the word *weed* is applied to any uncultivated plant in the field scorned by man, *condemning many beautiful native plants.*

"Weeds" should not always be considered man's enemies. Common "weeds" restore eroded land, help hold water during dry spells, and add beauty to our landscape. Some "weeds" have medicinal purposes, and others are valuable as food. A "weed" may be a nuisance to some people but desirable to others. Clover is a "weed" in gardens, but farmers plant certain kinds as food for their animals. Dandelions in lawns are annoying, but in some areas they are planted as a garden vegetable. To illustrate how the concept of "weeds" can be a frame of mind: a Japanese visitor to America related on his return to Japan that what impressed him most in the United States was the beautiful sight of the early spring fields covered with the bright yellow flowers called Dandelions!

Garden and roadside "weeds" have a vital function in the scheme of nature. Unfortunately, chemical herbicide operations and overzealous cutting prevent them from performing their valuable service of improving soil structure, retaining soil moisture, drawing minerals up from the subsoil, and adding these nutrients to the surface. More and more farmers are finding "weeds" beneficial when used as companion plants between rows of cultivated crops. Highway departments are reevaluating controlled "weeds" and wildflowers as an economical way to maintain roadways and are finding controlled "weed" growth a helpful factor in ecology. It is true that a few "weeds" such as Poison Ivy and Canadian Thistle need to be controlled. *Most "weeds," however, are valuable and useful to the environment* and add great beauty to the landscape.

There is a constructive relationship between soil and deep-feeding "weeds." Ragweed, unfortunately a nuisance to hay-fever sufferers, is one of the best soil-improving "weeds" in the United States, as it will grow in and improve almost any soil. Ragweed will also detoxify contaminated soil. It holds disturbed soil in place until other plants become established and then, surprisingly, the Ragweed is displaced. Plants of the Goldenrod, Milkweed, and Sunflower families, and particularly Sweet Clover, are also important soil builders.

Not only do "weeds" improve the soil for man, they also supply food for birds and wild animals and forage for farm animals.

But their worthwhile contributions do not end with these vital ecological factors. After the pods have released their seeds, the containers—the dried pods—remain in a variety of shapes and textures to provide yet another gift of nature, aesthetic beauty to feed man's soul.

Help dispel the myth that "weeds" have little or no value. Enjoy "weeds" for their functional place in ecology, for their colorful blooms in the landscape, and finally for their pods, the seed-holding containers of the plants, which are the wildflowers and weeds in their final beauty.

WHAT IS A POD?

It is common to associate the word *pod* with seed containers such as peas, beans, Milkweed pods, and the like, but for want of a better word and for purposes of this book, *any container or vessel holding seeds is considered a pod.*

Botanists may take exception to this generalization of the meaning of *pod.* The intention of this guide, however, is to identify the dry seed container, the end of growth; to identify the shape that bears little resemblance to the bloom. We are interested in *this final stage of the plant,* be it called pod, capsule, flower head, cup, bur, fruit, or nut.

How wonderful it is to discover that it is not necessary to lose interest when flowers fade! Becoming aware of the seed containers with their new shapes, colors, and textures extends the opportunity to observe the growing cycle of wild plants.

WHERE TO FIND DRIED PODS

Most wild plants need certain growing conditions and will generally be found in similar situations from area to area, in sun, shade, sand, or on shorelines. At times, however, unexpected species are found in unexpected places—at *the edge* of certain areas, especially along fencerows, under electrical lines, or at the edge of the woods, where birds perch and drop seeds.

Other locations not to be overlooked:

• Railroad rights-of-way are good hunting grounds, as passing trains swirl dropped seeds about and the open area encourages growth.

• Disturbed land generally contains the coarser, more vigorous types of wild growth. Seeds can remain dormant for years and upon being "turned up" will thrive again.

• Prairies, meadows, and grasslands produce more delicate flowers— flowers that prefer undisturbed conditions.

• Wetlands are good areas for finding interesting species because animals and machinery do not find them accessible.

• Hillsides too steep to plow or for animals to graze on are also areas worth investigating.

• Fencerows where plants are left undisturbed may be a source of unexpected specimens.

• Land where the soil is too poor for agriculture may still yield interesting wildflowers and weeds.

• Around abandoned farmhouses, barns, factory-industrial areas, and vacant lots weeds thrive.

Do not hesitate to return to places you have previously visited, as each year new varieties spring up to enrich your collecting pleasure.

Weeds are found everywhere—even at your own doorstep!

WHEN TO PICK PODS

To harvest pods at their peak beauty, take several trips to the area to acquaint yourself with the habit of the flower. It does make collecting easier to see a flower in some area and know what the pod will look like for later gathering. (I keep a small notebook in my pocket or car and earmark locations throughout the flowering season for specific pods I will want at picking time. It is also a good idea to keep some large plastic bags, ties, cutting shears, hair spray, and boots in the trunk of your car in case you spot a collectible pod. For extra enjoyment, I also use a magnifying glass.)

Generally, *the best time to pick is when the leaves begin to wither or have begun to drop and the stems are turning brown*—after maturity, but before frost. (By the time the pods are dry enough to pick, the wind has usually taken care of scattering the seeds.) Colors will be at their brightest. If pods are left in the field too long, the colors become drab, faded, or mottled, and the pods can be torn by winds and rain. For example, the Prairie Dock flower head keeps its sculptured rosebud shape, and the dramatic leaf is a gorgeous leathery plume shape of burnished bronze if picked in mid-September or before it is nipped by frost. If weathered too long, however, the head becomes ragged, and the leaf turns spotty and a dull dark brown. The Whorled Milkweed pod loses its delicate fawn color, develops black spots, and is easily shredded by the wind.

The delightful part about collecting dried pods is that no involved process is needed to preserve the dried materials except picking at the proper time and careful storage of the collection in a dry place until the pods are ready to be used.

POD PICKING,
SOME DOS AND DON'TS

- Learn the names of the protected wildflowers in your area and protect them. Flowers such as Buttercups, Violets, Asters, and Goldenrods can be gathered without fear of extermination. Trilliums, Orchids (Lady Slipper), Jack-in-the-Pulpits, and Gentian plants are destroyed by picking. *Do not* collect these plants. There is a belated movement to protect endangered plants. Learn these species and admire them in place, but resist the temptation to take seed or pod.
- Do not pick a pod *if you cannot identify it.* It might be a protected plant.
- Be considerate of the owners of the land. Collecting should be restricted to areas where permission has been granted, or where plants are being destroyed by construction, road work, and so on.
- Pick larger bouquets of pods from among the coarser flowers, as they are usually in less danger of extinction than the delicate ones and often are better for decorative purposes.
- Do not dig up plants for transplanting to your home garden. Generally, growing wildflowers in a domestic garden is difficult. If you must have a species, plant some of the seeds you collect or order plants from a wildflower nursery.
- Do not uproot plants when picking. Cut or break the stems carefully.
- Always leave some of each species intact for the birds and animals. *Never strip an area.*
- Pick pods after they have dried in the field, because that is usually when they are at their premium beauty and color and *seeds have had a chance to drop.*
- Shake out the remaining seeds from the pods as you pick them to propagate for future picking.
- Do not pick indiscriminately and wastefully.
- Strip off the dried leaves, and hang pods upside down in bunches in a cool, dry place, a barn or garage, to finish drying and to keep heads erect.
- Do not pick dull white or creamy waxlike berries from plants *with leaves*

in groups of three. These are Poison Ivy. Poison Oak and Poison Sumac have similar berries. Poison Ivy grows in patches on the ground or as a climbing shrub. The berries grow in a tight cluster. Poison Oak is similar but grows as a bush (found in the South and on the Pacific coast). The related Poison Sumac is a shrub or small tree with 7 to 13 *untoothed leaflets* on an axis and berries hanging in loose clusters. (Poison Sumac should not be confused with the harmless Smooth Sumac or Staghorn Sumac, both of which have *toothed leaflets* and *berries in fuzzy red clusters standing upright at the ends of branches.*) All the Poison Ivy group have colorful fall foliage. Learn to recognize them and leave them alone.

• *Do what you can in your community to express the need to save our wild plants.* Don't let your roadsides be cut back too far and chemically sprayed. The mowing of waysides alters the natural balance of our vegetation. With fewer wildflowers left to bloom, our colorful butterflies, grass fowl, and many small animals are becoming victims of man's encroachment on the natural environment. Explain to local highway maintenance officials and city public works officials the ecological values of wild vegetation. Encourage them to let wildflowers bloom. Patches of Chicory, Black-eyed Susan, Spotted Knapweed, Queen Anne's Lace, Goldenrod, and Frost Aster add color and softness to our countryside and expressway areas, and later in the season, the beauty of the pods will enhance the landscape.

CREATING ARRANGEMENTS WITH PODS AND DRIED MATERIALS

Bouquets of dried wildflowers add visual warmth and beauty to the home. They can be as colorful as fresh flowers, and they last indefinitely. An almost endless assortment of size, shape, and texture of pods is available, which allows for creative self-expression in producing a wonderful variety of arrangements.

Collecting pods is a year-round pastime. It almost becomes a game to be on the lookout for new pods and exciting accessories to include with the pods. Working with *dried* materials has the advantage of ready supplies at your fingertips from autumn to spring.

Not only are dried pods and flowers to be considered, but also dried leaves, ferns, twigs, driftwood, berries, and fruits. Anything that can be *recycled* from nature and is pleasing to the eye should be saved and used.

Be creative in designing arrangements with pods. They can be unusual or simple, formal or casual, a huge mixed grouping or a bouquet with just one or two types of pods. Often a single spray or branch in a unique container will become a work of art. Some pods are so interesting in their shape and texture that additional pods would detract from their beauty.

Containers and the way the pods are arranged set the mood for the grouping—formal in a china vase, or a random bunch stuck in a jug with carefree abandon, an asymmetrical or balanced arrangement, ball-shaped or pointed, tall or short, rugged or delicate, miniature to floor-size; pods and containers almost suggest arrangements themselves.

For added color in your arrangements, preserve some of the wildflowers and weeds in bloom, remove the leaves if you wish, tie them in bunches, and hang them upside down to dry in a ventilated, dark, dry place. (For example, various species of Goldenrod retain their beautiful golden yellow; Smooth Sumac picked in summer remains a rich, deep red; Upland Boneset turns a lovely soft lime green.) Allow flowers to dry several weeks to harden stems.

Also, in your own garden, you may wish to grow colorful flowers especially cultivated for drying, such as Honesty (Silver Dollar), Statice, Globe Amaranth, Helichrysum (Strawflowers), Everlasting Flowers, Acroclinium, and Job's Tears. All will add lovely bright colors to your bouquets.

Berries can also be used to add beautiful colors to your arrangements. Be sure they are well hardened, or they will shrivel when brought indoors. Allow rose hips to remain on the plant until crisp October weather for rich red-colored fruits. Horse-nettle has a lovely yellow-green cherry-size fruit. The Carrion-flower vine has dramatic blue-black clusters of tightly bunched berries on its tendril-like branches. Some Wild Asparagus has deep orange berries all over the straw-yellow, wispy branches. All are good accent colors. Bittersweet, a very popular dried bouquet berry, is protected in some states but can be purchased·at florists and markets.

You can enjoy the natural, subtle, earth-tone colorings of pods, or you may want to spray or dye them for specific colors to complement your decor or to highlight a holiday theme. (See page 41 for a Fourth of July suggestion.)

Shake fuzzy heads of dried flowers to avoid messy shedding and to obtain strawlike flowers, or spray with lacquer or hair spray if you want to add a soft, fluffy touch to your floral piece. Spray Yellow Goat's-beard with hair spray *before* it is picked to protect the fragile seed head. (Yellow Goat's-beard is one exception to the rule of shaking out the seeds!)

Containers create the theme for your floral pieces. Don't hesitate to use all forms of containers: vases, contemporary pottery, ordinary clay pots, wine bottles, kitchen household items, wooden bowls, cups, beer mugs, pitchers, teapots, butter crocks, milk jugs, snifter goblets, buckets, berry baskets, all sizes and shapes of fiber and woven baskets, brass, pewter, copper, glass, antique holders, spice boxes, grinders, graters, decanters, pharmaceutical bottles—to mention a few.

Hold arrangements in place by filling containers with sand or crushed chicken wire to keep stems in position, or use needle holders in shallow containers. Brown or green styrofoam cut to fit and wired inside is especially good in open-weave baskets or in bowls. Secure the styrofoam to the bottom of other containers with florist's "stay-stuck" tape or sticky putty. All these products are available at florists or garden supply shops, and new ideas and suggestions are always available there.

Add special interest to arrangements. Use bare twigs such as Red Dogwood, Corkscrew Willow, or dark shiny branches of Choke Cherry for accent. You can also use field grasses and grains, sea grass, Cat-tail leaves, Feather Grass plumes, domestic bird tail feathers (such as rooster, turkey, goose, pheasant, and peacock), colorful dyed weeds, and bleached weeds; and you can wire stems onto pine cones, Wild Cucumber pods, nuts, acorns, and avocado seeds. Try dried artichokes and okra pods, Indian corn with husk leaves, corn tassels, cornhusk flowers, feather flowers in all sizes and colors, and unusual pods such as the Lotus pod or the Hawaiian bean imported from exotic lands. Gift and craft shops sell artificial insects, butterflies, birds, and fruit that add interest to bouquets. Driftwood, ribbon, and bows add finishing touches.

Dried-on-the-tree Oak leaves, naturally dried and curled Common Milkweed leaves and stalks, dried russet-colored fern leaves, lacy remains of the Elderberry and Dogwood clusters, and the dried Frost Aster and Sea Lavender are some good fill-in materials.

The photographs of dried arrangements on the following pages may suggest ideas for your own creative enjoyment. In many instances an interesting container and a single species have been arranged to emphasize the subtle beauty of the pod.

The names of Ruth M. Weber and Colleen J. Weiler are added below the arrangements that they designed. The rest of the arrangements were created by the author.

. . . JUST THE BEGINNING

This book is just the beginning of getting to know pods. There are many more lovely pods, too numerous to be included here; yet they deserve discovery. Listed below are other species that develop decorative pods that were not included at this time. Refer to a flower guide to become acquainted with the flower and then watch for the pod to mature.

Many pleasant surprises are in store and the personal identification adds challenge and the thrill of self-discovery to the hunt.

AGRIMONY—*Agrimonia*
AVENS, LONG-PLUMED PURPLE—*Geum triflorum*

BASIL, WILD—*Satureja vulgaris*
BEECHDROPS—*Epifagus virginiana*
BITTERSWEET, WILD—*Celastrus scandens*
BLUECURLS—*Trichostema dichotomum*
BONESET—*Eupatorium perfoliatum*
BUGLOSS, VIPER'S—*Echium vulgare*
BUTTERCUP, TALL or COMMON—*Ranunculus acris*

COHOSH, BLUE—*Caulophyllum thalictroides*
COREOPSIS, STIFF—*Coreopsis palmata*

ELECAMPANE—*Inula helenium*

FERN, SENSITIVE—*Onoclea sensibilis*
FLAX, BLUE—*Linum usitatissimum*

GARLIC, WILD—*Allium canadense*
GERMANDER—*Teucrium canadense*
GOLDENROD, BOG—*Solidago uliginosa*
GRASS-OF-PARNASSUS—*Parnassia glauca*

INDIAN-PIPE—*Monotropa uniflora*
INDIAN-TOBACCO—*Lobelia inflata*

JERUSALEM ARTICHOKE—*Helianthus tuberosus*
JIMSONWEED—*Datura stramonium*

KNAPWEED, BROWN—*Centaurea jacea*
KNAPWEED, TYROL—*Centaurea vochinensis*

LEATHER-FLOWER—*Clematis viorna*
LEEK, WILD—*Allium tricoccum*
LILY, CANADA—*Lilium canadense*
LOOSESTRIFE, FRINGED—*Lysimachia ciliata*
LOOSESTRIFE, YELLOW—*Lysimachia terrestris*
LUPINE, WILD—*Lupinus perennis*

MALLOW, SWAMP ROSE—*Hibiscus palustris*
MERCURY, THREE-SEEDED—*Acalypha rhomboidea*
MONKEY-FLOWER, SQUARE-STEMMED—*Mimulus ringens*
MULLEIN, MOTH—*Verbascum blattaria*
MUSTARD, TOWER—*Arabis glabra*
MUSTARD, TUMBLE—*Sisymbrium altissimum*

OX-EYE—*Heliopsis helianthoides*

PARTRIDGE-PEA—*Cassia fasciculata*
PEA, EVERLASTING—*Lathyrus latifolius*
PENNYROYAL, AMERICAN—*Hedeoma pulegioides*
PIMPERNEL, YELLOW—*Taenidia integerrima*
PINEWEED—*Hypericum gentianoides*
PINWEED—*Lechea intermedia*
PIPSISSEWA—*Chimaphila umbellata*

SAXIFRAGE, SWAMP—*Saxifraga pensylvanica*
SEA-LAVENDER—*Limonium nashii*
SEEDBOX—*Ludwigia alternifolia*
SKULLCAP, HEART-LEAVED—*Scutellaria ovata*
SNEEZEWEED—*Helenium autumnale*
STEEPLEBUSH—*Spiraea tomentosa*
STONECROP, DITCH—*Penthorum sedoides*
SUMAC, SMOOTH—*Rhus glabra*

VIRGIN'S-BOWER—*Clematis virginiana*

YUCCA—*Yucca filamentosa*

COLOR GUIDE
FLOWERS, PODS, ARRANGEMENTS

There are variations in almost all species. Some of the most characteristic ones have been selected.

The pods in a specific family are quite similar; for example, there are many types of Goldenrod that vary in silhouette—flat-top, plumelike, elm-shaped, spear-headed—yet all are similar in having fuzzy-headed pods and strawlike florets when the hairy seeds blow away.

The same is true for the Asters, Milkweeds, Mustards, Cresses, Bonesets, and so on. Through this collection of pods it is hoped that the reader will be able to recognize the *species* and identify the variations from more complete flower guides.

FLOWERS

Bloom: Apr.–June 6–20 in. tall
Flowers: White, small, 4 petals in racemes
Leaves: Alternate, toothed, clasp stems,
       branchy
Found: Disturbed areas, pastures;
       common

FIELD PENNYCRESS
Thlaspi arvense

PODS (early May–June)
If pods get too dry, they fall off and only
skeleton membranes remain. Pods are
large, deeply notched, flat, and strikingly
light-straw-colored. Spray them to prevent
dropping.
• Use this good pod as a filler or for its
light color; the skeletons add a wispy effect to arrangements.

Field Pennycress, apples, pine cones

27

SUN

FLOWERS

Bloom: Apr.–Aug. 1–2 ft. tall
Flowers: Bright yellow, 4 petals, bloom
up spikes
Leaves: Lower ones cut into 5 divisions
with large terminal lobes
Found: Roadsides, fields, disturbed
areas, moist areas; common

YELLOW ROCKET
Barbarea vulgaris

PODS (early June on)
Yellow Rocket (Winter Cress) may be
picked all summer as flowers or as pods;
hang to dry. The empty, long, pale-wheat-
colored pods look like transparent pine
needles climbing the stems.
• This fragile, spiny, bleached pod is a re-
freshing addition to mixed arrangements
but is also quite sleek when used alone.

Yellow Rocket, dried blooms and pods
(Weber)

28

FLOWERS

Bloom: May–Sept. 3–18 in. tall
Flowers: Greenish-yellow cone-shaped head, no rays
Leaves: Alternate, fernlike, crushed leaves smell like pineapple
Found: Roadsides, barnyards, paths, disturbed areas

PINEAPPLE WEED
Matricaria matricarioides

PODS (midsummer on)

Allow Pineapple Weed to dry in the field. When the seeds fall, sand-colored, shallow, cuplike pods with pointed cone centers remain (resembling miniature Mexican sombreros). The lovely pineapple scent lingers with the pods.

• This plant was included because its diminutive size lends itself unusually well to use in small containers.

Pineapple Weed, Rose hips (Weber)

FLOWERS
Bloom: May–Oct. 4–16 in. tall
Flowers: Pinkish gray, furry elongated
            heads
Leaves: Palmate, narrow, alternate
Found: Roadsides, dry fields; common

RABBIT'S-FOOT CLOVER
Trifolium arvense

PODS (early summer on)
Surprisingly, the fuzzy head of Rabbit's-foot Clover does not disintegrate, nor does it lose its lovely pastel color when dried. Although fragile in appearance, it preserves well. Since the very fine stems tend to become brittle, arrange carefully.
• The gray-pink cast of this wildflower is beautiful in miniature bouquets as it adds a soft touch.

Rabbit's-foot Clover

FLOWERS

Bloom: May–Oct. 8–24 in. tall
Flowers: White, bushy, encircle lower part
of thimble heads
Leaves: Basal, long, slender, grasslike
Found: Disturbed areas, very common in
grassland

ENGLISH PLANTAIN
Plantago lanceolata

PODS (mid-June on)

This plant is a familiar lawn and wayside sight all summer long. The brown seeds are crowded at the top of the erect stem, giving the head a scrubby, elongated, thimblelike look.

• These tubular pod heads are rougher-looking than most small pods and add a grainy interest to small mixed arrangements.

English Plantain, Whorled Milkweed
(Weiler)

31

FLOWERS

Bloom: May–Sept. 6–18 in. tall
Flowers: White to greenish, in dense
racemes at top of branches
Leaves: Arrow-shaped, alternate, base
ones clasp stems
Found: Meadows, roadsides, disturbed
areas; common

FIELD PEPPERGRASS
Lepidium campestre

PODS (end of June on)
Field Peppergrass pods (also known as
Cow Cress) may be found all summer.
Pods still with seeds look like miniature
spoons clinging to the stem. When seeds
drop, a transparent, parchment-white,
pointed-tip skeleton remains.
• Pods with seeds give a dotted look to
stem and make excellent fillers in arrangements; for a more delicate feeling use
seedless skeletons.

Field Peppergrass, Black-eyed Susan
(Weber)

FLOWERS
Bloom: April–June   1–3 ft. tall
Flowers: White, climb up loose racemes
Leaves: Heart-shaped, sharp-toothed,
alternate, garlic odor
Found: Roadsides, disturbed areas,
open woods; common

GARLIC MUSTARD
Alliaria officinalis

PODS (early July–Aug.)
Pick some pods about the time blooming
stops and pods are still full and green.
Hang to dry for lovely green color and for
upturned, slender bean-pod shape. Later
in summer pick the light-colored, seedless,
transparent skeletons but do so before
they become too dry and tattered.
• The dried, spinelike, wispy appearance
is lovely in arrangements, but handle carefully as dried pods crumble easily.

Sprayed Garlic Mustard, Goldenrod,
sprayed Goldenrod insect galls,
False Boneset (Weiler)

FLOWERS

Bloom: June–Oct.   1–3 ft. tall
Flowers: Light yellow with orange palates,
resemble Snapdragons
Leaves: Very narrow, linear, numerous
Found: Fields, disturbed areas,
roadsides; very common

BUTTER-AND-EGGS
Linaria vulgaris

PODS (midsummer to late Sept.)
Some years pods are quite ragged and colorless. Other years nice firm, straw-colored, upturned bells climb the stems. They tend to fall to the ground when dry, so pick as soon as pods develop.
• Since pods and spikes are rather delicate, they make attractive smaller arrangements.

Butter-and-eggs, dried Wild Asparagus fern (Weber)

FLOWERS
Bloom: June–Sept. 8–30 in. tall
Flowers: Pink or white, tiny, at top of
inflated bladders
Leaves: Linear, opposite, on stems with
dark sticky areas
Found: Dry sandy soil, disturbed areas;
common

SLEEPY CATCHFLY
Silene antirrhina

PODS (July–Sept.)
This fragile-appearing pod seems to thrive
in very rough or sandy areas. It is almost
overlooked because of its diminutive size
and its concealment by more robust
plants. Let specimens dry in the field.
• Since it is such a dainty pod, it is better
used in small bunches when included in
mixed bouquets; it is also excellent in miniatures.

Sleepy Catchfly

35

ALSIKE CLOVER
Trifolium hybridum

FLOWERS
Bloom: May–Sept. 1–2-ft. runners
Flowers: Creamy white centers, pink
 edges
Leaves: Palmate, 3 oval leaflets
Found: Fields, disturbed areas,
 roadsides

PODS (July into fall)
If picked early, these pods have creamy to pale green centers and are edged with tinges of russet brown. Hang to dry with leaves for green contrast. Picked later, pods become brown and ragged-looking.
• Small Alsike Clover is very attractive in early American or miniature arrangements.

Alsike Clover

FLOWERS

Bloom: June–Sept. 6–18 in. tall
Flowers: Green, long, tightly flowered
      spike heads
Leaves: Broad, egg-shaped, basal,
      strongly ribbed lengthwise
Found: Yards, paths, roadsides; like
      shade and sun; very common

COMMON PLANTAIN
Plantago major

PODS (July on)

Good long spikes of Common Plantain are hard to find at roadsides, as they are continually cut down. Matured plants turn deep brown with tightly packed seeds around the entire stem, resulting in a pencil-size, textured cylindrical shape.
• This pod gives an interesting "stuck-in" or "jabbed" appearance to dried bouquets.

Common Plantain, Strawflowers

FLOWERS

Bloom: June–Sept. 8–24 in. tall
Flowers: Yellow-orange, several atop
 each stem
Leaves: Basal rosette, oblong, plants
 very hairy
Found: Fields, roadsides, disturbed
 areas

ORANGE HAWKWEED
Hieracium aurantiacum

PODS (July–Oct.)

Orange Hawkweed blooms and pods throughout the summer. If flowers are picked, some acquire fluffy, light green-gray seedpods and others remain dried rays of deep orange. Hang to preserve and harden, as heads seem too heavy for the soft, hairy stems.

• There is an unusual color quality about this pod that makes it visually more pleasing when used sparingly or alone.

Orange Hawkweed

FLOWERS

Bloom: June–Sept. 2–6 ft. tall
Flowers: Deep magenta-purple, upright in
      bracts, covered with yellow
      spines
Leaves: Prickly, long, pale, irregularly
      lobed, on prickly stems
Found: Roadsides, clearings, pastures;
      common

BULL THISTLE
Cirsium vulgare

PODS (July on)

Though this entire plant is spiny and hard
to handle, the flowers and pods are worth
collecting and may be picked all summer.
Flowers hung upside down will dry to pale
lavender heads with soft green bracts; mature pods are sandy beige and spiny in appearance.

• Surprisingly, these large, rugged pods
lend themselves well to subtle oriental arrangements.

Bull Thistle, dried green Garlic Mustard

FLOWERS

Bloom: June–Oct. 4–12 in. tall
Flowers: Yellowish green turning brown-red, tiny, on branching spikes
Leaves: Long, arrowhead-shaped, with spreading lobes
Found: Roadsides, fields, acid soil.

SHEEP SORREL
Rumex acetosella

PODS (July on)

This plant is rather inconspicuous while green but gradually turns ruddy red-browns, and it has a pleasing, lacy, upright appearance. It is a good pod to collect for use in small and miniature containers; the stems are so fragile, the pods so minute, and the whole plant so flushed in variations of "sorrel" brown.

● It is charming used alone or used in bunches in mixed bouquets.

Sheep Sorrel (Weiler)

FLOWERS

Bloom: June–Oct. 1–3 ft. tall
Flowers: Medium yellow, square-tipped
      rays, long pointed bracts; close
      at midday
Leaves: Long, grasslike, light green,
      clasp smooth stems
Found: Fields, roadsides, disturbed
      areas; common

YELLOW GOAT'S-BEARD

Tragopogon pratensis

PODS (July on)

Yellow Goat's-beard continues to burst
into large, fluffy seed heads all summer
long. Spray *before picking* and keep heads
apart when gathering. Try spraying in a
color for variety. They are fragile, but the
"see-through" quality is lovely.

• In fall, pick some of the dry bare stems;
they have an unusual contour and add interesting thrusts to arrangements.

Sprayed Yellow Goat's-beard, sprayed
Narrow-leaved Cat-tail

41

FLOWERS

Bloom: June–Oct. 1–3 ft. tall
Flowers: Bright yellow rays, cone-shaped,
        brown centers
Leaves: Oblong, hairy, on rough stems
Found: Fields, disturbed areas,
       roadsides

BLACK-EYED SUSAN
Rudbeckia hirta

PODS (mid-July–Oct.)

These pods are noticeable by the almost black cone-shaped heads erect on slender, hairy stems. Pick early for satiny dark heads and hang to dry if stems are still green.

• These pods adapt well to small and medium-size bouquets and add nice sharp, dark accents.

Black-eyed Susan, Italian Starflowers, Wheat
(Weber)

FLOWERS

Bloom: July–Sept. 1–5 ft. tall
Flowers: Golden yellow, in plumelike
　　　　clusters
Leaves: Lance-shaped, sharp-toothed,
　　　　crowded
Found: Meadows, clearings, roadsides;
　　　　common

CANADA GOLDENROD
Solidago canadensis

PODS (July–Sept.)
FOR FLOWER HEADS: Pick as soon as blooms appear and hang to dry for gorgeous variations of yellow; pick later in Sept. for dried fuzzy tops and in Oct.–Nov. for seedless strawflower heads. (There are about 70 species of Goldenrod with differences in shapes but similarities in pods.)
• Use Goldenrod in any arrangement from colonial to contemporary.

Canada Goldenrod
(Weiler)

43

FLOWERS
Bloom: April–Sept. 8–18 in. tall
Flowers: White, tiny, on end of stems
Leaves: Basal, coarsely lobed, stem
leaves lance-shaped
Found: Everywhere; roadsides,
disturbed areas; very common

SHEPHERD'S PURSE
Capsella bursa-pastoris

PODS (throughout the summer)
These common roadside wildflowers are easily identified by the tiny, heart-shaped pods (shepherd's purses) that develop up the stem. If picked before "purses" become too dry, they will not shed as readily. Without seeds the transparent, pointed membranes add a flair to the stem.
• The pale sand color and the airy, patterned branches lighten and brighten any bouquet.

Shepherd's Purse, dyed Italian
Starflowers (Weber)

FLOWERS

Bloom: June–Oct. 1–4 ft. tall
Flowers: Reddish purple, nodding,
flat-domed, with reflexed bracts
Leaves: Very spiny, alternate, deeply
lobed
Found: Fields, roadsides, disturbed
areas

NODDING THISTLE
Carduus nutans

PODS (July on)

Blooms may be picked all summer to retain pale lavender color and hung upside down to dry. However, naturally dried pods are especially nice! They become beautiful, large, sun-bleached, and strawflower-like with huge, butter-yellow, velvet centers.

• There is a tendency for the dried pod to "nod" down. Remove head from natural stem and insert wire stem bending in direction wanted.

Nodding Thistle, miscellaneous
grasses

POOR-MAN'S PEPPER
Lepidium virginicum

FLOWERS
Bloom: June–Nov. 6–24 in. tall
Flowers: White, small, at tip of racemes, ripe seedpods below
Leaves: Basal, oval, deeply toothed, upper leaves linear
Found: Fields, roadsides, disturbed areas; common

PODS (July on)
These pods may be found all summer long. For variety, pick some in bloom and dry for white-tipped stems and pastel green pods. If left in field too long, seeds drop, and transparent, miniature, leaf-shaped skeletons march up the stem.

• With or without seeds, Poor-man's Pepper makes an attractive dried shape and its color gives a light, airy appearance to any bouquet.

Poor-man's Pepper

FLOWERS

Bloom: June–Aug. 1–3 ft. tall
Flowers: White rays, yellow center disks
Leaves: Basal leaves much lobed, upper leaves narrower and smaller
Found: Old fields, roadsides; very common

WHITE DAISY
Chrysanthemum leucanthemum

PODS (mid-July–Aug.)
Maturing in various shades of gray, pods look like curled-in strawflowers at end of the straight stems. For lighter color, pick some as soon as rays dry and center disks turn ocher yellow. Hang to dry.
• These pods add sharp upright thrusts to bouquets and are nice used in small containers.

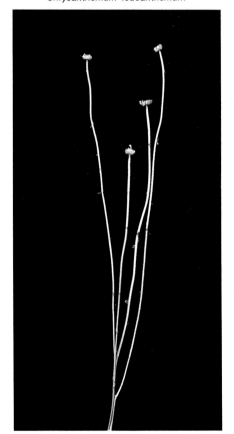

White Daisy on cactus skeleton

47

FLOWERS
Bloom: July–Aug. 1–2½ ft. tall
Flowers: White, in branching clusters
Leaves: Narrow, smooth, pointed
Found: Dry woods, fields, thickets

MOUNTAIN-MINT
Pycnanthemum virginianum

PODS (mid-July on)
To help identify, crush pod heads for the mint fragrance. Mountain-mint has long, fragile stems with many upturned, fine branches near the top crowned with flat-topped heads. Pick early for soft silver gray; gather later for gunmetal gray.
• These button-top pods on light stems look good in most arrangements; they are also interesting pods when sprayed a bright color.

Sprayed Mountain-mint

FLOWERS

Bloom: June–Aug. 1–4 ft. tall
Flowers: Yellow-green clusters climb
       spikes
Leaves: Long, lanceolate, with wavy
       edges
Found: Roadsides, pastures, disturbed
       areas; very common

CURLED DOCK
Rumex crispus

PODS (mid-July on)

Curled Dock may be picked all summer for its wide range of colors. It is yellow in bloom and gradually turns green, honey-toned, rich red brown, and later very dark brown.

• The husky, full look of these spikes is desirable for large arrangements and is especially pleasing used as background in one-sided bouquets.

Curled Dock, dried Goldenrod, Aster, Rose hips, Figwort, Whorled Milkweed, Strawflowers, dried leaves (Weber)

FLOWERS
Bloom: June–Sept. 1–2 ft. tall
Flowers: White, at end of long racemes
Leaves: Narrow, alternate, hoary down
on leaves and stems
Found: Roadsides, disturbed areas,
meadows

HOARY ALYSSUM
Berteroa incana

PODS (mid-July on)
The pods swollen with seeds are oblong
with short, pointed tips. When seeds dis-
perse, transparent, tear-shaped skeletons
climb the entire stem. Dry some green
pods and pick some empty pods for variety.
• These airy, pale pods add a soft, linear
touch to dry arrangements.

Hoary Alyssum, Statice (Weiler)

50

FLOWERS

Bloom: June–Sept. 1–3 ft. tall
Flowers: White, lacy, flat-top umbels,
deep purple center florets
Leaves: Carrotlike, finely divided
Found: Roadsides, disturbed areas,
fields; common

QUEEN ANNE'S LACE
Daucus carota

PODS (mid-July on)
Queen Anne's Lace is a very beautiful and familiar wildflower in summer bouquets, and its pods are equally attractive. As the flower matures it curls inward, giving the appearance of a little bird's nest. Under certain weather conditions and in some areas the heads of dried flowers remain flat and open. Pick early for good color.
• Both shapes are decorative and add light, airy touches to arrangements.

Queen Anne's Lace, Corkscrew Willow (Weiler)

51

FLOWERS

Bloom: June–Oct. 1–3 ft. tall
Flowers: White, very small, in terminal
  clusters
Leaves: Narrow, light green, lower leaves
  alternate, upper leaves in whorls
  at base of inflorescence
Found: Roadsides, open woods, dry
  sandy areas

FLOWERING SPURGE
Euphorbia corollata

PODS (mid-July on)
This plant eventually turns rosy shades
and looks wispy. Pick when in red tones
and dry upside down. Pods are not dramatic; it is the shape of the entire plant and
the ruddy color of drying leaves that is desirable for arrangements.
• Use Flowering Spurge for a color touch.
It is particularly adaptable to dainty, femi-
nine containers.

Flowering Spurge, sprayed Goldenrod
(Weber)

FLOWERS

Bloom: July–Sept. 1–4 ft. tall
Flowers: None; spores are fruiting heads
Leaves: None; sterile stems, hollow,
toothed sheaths at stem joints
Found: Wet sandy soil, gravelly soil,
roadside lowlands

HORSETAIL
Equisetum hyemale

PODS (mid-July on)
Horsetails (Scouring Rushes) are lovely soft green, almost bamboolike shoots with black rings at the joints. They mature throughout the summer. Pick when size appeals and hang to dry in a dark, dry place to harden and to avoid fading.
• These decorative stems of soft color add lovely directional lines to any arrangement but especially to those of oriental design.

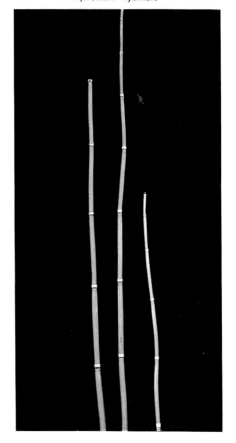

Horsetail, Bull Thistle

53

FLOWERS

Bloom: May–Sept. 2–5 ft. tall
Flowers: Deep yellow, in umbels
Leaves: Pinnately compound,
        sawtoothed, on grooved sturdy
        stems
Found: Roadsides, disturbed areas,
       meadows; common

WILD PARSNIP
Pastinaca sativa

PODS (end of July–Sept.)
This is a plant that quickly becomes scraggly-looking with loss of seeds, so keep watching for leaves to dry and pods to mature. Gather while full-headed if possible. Spraying is suggested to prevent shedding.

• The seeded umbel heads add a lacy look to any bouquet, and huge bunches of branched Wild Parsnip provide a lovely, open feeling.

Wild Parsnip, stained Heal-all, Black-eyed Susan, Yarrow, dried leaves

FLOWERS
Bloom: July–Sept. 1–2 ft. tall
Flowers: Pearly white, globular, with
yellow-brown centers
Leaves: Long, linear, wooly white leaves
and stems
Found: Dry sandy soil, hillsides, woods;
in patches

PEARLY EVERLASTING
Anaphalis margaritacea

PODS (end of July–Sept.)
For a brighter pod, pick as flowers and
hang to dry. The pods look much like the
flowers and remain pearly white with dark
gold-brown centers. Stems and leaves are
very light gray.
• This is one of the prettiest of the dried
materials and adds a light, cheerful touch
to any bouquet.

Pearly Everlasting, garden Yarrow, Tansy,
Curled Dock, Silver Dollars, Statice, Strawflowers, Foxtail grass, Baby's Breath, Arte-
mesia (Weiler)

BOUNCING BET
Saponaria officinalis

FLOWERS
Bloom: July–Sept. 1–2 ft. tall
Flowers: Pink or white, enclosed in
tubular calyxes, clustered
Leaves: Opposite, smooth, lanceolate,
stems thick-jointed
Found: Railroad banks, roadsides,
disturbed areas

PODS (end of July on)
This plant (also known as Soapwort) is one
that should be watched and picked as
soon as it goes to pod. Later the pods become shabby-looking and dull in color.
• A compact head on a sturdy stem makes
this pod easy to handle; it is good used as
a filler and lends weight to bouquets.

Bouncing Bet, Pearly Everlasting, Silver
Dollars, Velvet-leaf

FLOWERS

Bloom: July–Sept. 1½–3 ft. tall
Flowers: Yellow-green, tiny, in long
upright clusters
Leaves: Lanceolate, pale, white felt
beneath
Found: Dry open soil, prairies,
roadsides, disturbed areas

WHITE SAGE
Artemisia ludoviciana

PODS (end of July on)

White Sage (Western Mugwort) is easily
sighted by the patches of silver-white
plants swaying in the green fields. It is a
statuesque, upright species with beautiful
leaves and tapering pod heads; the entire
plant is almost an alabaster white.

• There isn't a lovelier, muted white, vertical accent than White Sage for mixed
color arrangements.

White Sage, dried Wild Marjoram,
Statice

FLOWERS

Bloom: July–Sept. 1–3 ft. tall
Flowers: White to pale violet with dark
spots, in clusters
Leaves: Opposite, heart-shaped, lighter
green or gray underneath, hairy
Found: Disturbed areas, roadsides;
common

CATNIP
Nepeta cataria

PODS (Aug.–Sept.)

Pick Catnip while it is still in bloom and the head is soft gray-white. Hang to dry. Picked later, it becomes a dusty, colorless gray.

• This is a worthwhile flower pod to collect for the gray contrast it gives to fall dried materials and is an especially good addition to Williamsburg arrangements.

Catnip, Early False Indigo, graden Yarrow, Curled Dock, Evening-primrose, Spotted Knapweed, Wheat (Weiler)

58

FLOWERS

Bloom: May–July 2–4 ft. tall
Flowers: White, inflated, tubular, with two lips
Leaves: In pairs, fine-toothed, stalkless
Found: Fields, open woods, prairies

FOXGLOVE BEARDTONGUE
Penstemon digitalis

PODS (Aug.–Sept.)
Beardtongue is another long-stem species that has graceful, branched pod heads. The pods, a warm nutmeg color, are pear-shaped, sitting in bract nests. Allow the pods to brown in the field.
• Show off the nicely detailed design in an uncluttered arrangement, so that the form may be appreciated.

Foxglove Beardtongue, grass leaves

HEAL-ALL
Prunella vulgaris

FLOWERS
Bloom: June –Sept. 3 –20 in. tall
Flowers: Violet-blue, upright cylindrical
        heads
Leaves: Opposite, oblong, slightly
        toothed
Found: Disturbed areas, roadsides,
       grasslands; common

PODS (Aug.–Sept.)
Heal-all (or Selfheal) may be found low and creeping or over 1 ft. tall. On branched stems, pods look like dried, stacked flowers, very trim and textured. For amber browns pick as soon as blooming ceases, but gather later for dark walnut browns.
• This sturdy, compact, cylindrical pod fits into any arrangement.

Heal-all, Purple Loosestrife, Foxtail grass

FLOWERS

Bloom: July–Sept. 1–2 ft. tall
Flowers: Pink on elongated gray-green heads
Leaves: Fine, pinnate, on long wiry stems
Found: Dry prairies, open woods

PURPLE PRAIRIE CLOVER
Petalostemum purpureum

PODS (Aug.–Sept.)

The flowers may be dried for the soft lavender-colored fringe and pale gray-green thimble heads. White Prairie Clover, *Petalostemum candidum,* is similar but has white flowers. Spray to prevent the pods from bursting open and producing disheveled, cottony heads.

● The thimblelike head on a spindly stem makes an appealing pod for use in small bouquets.

Purple Prairie Clover, Rush grass (Weber)

FLOWERS

Bloom: June–Aug. 8–20 in. tall
Flowers: Deep magenta, white spotted petals, long bristly bracts
Leaves: Linear, very fine, opposite
Found: Disturbed areas, poor dry soil, roadsides

DEPTFORD PINK
Dianthus armeria

PODS (Aug.–Sept.)

These small, very slender plants are easily overlooked in the field. On erect stems their off-white, curled, fine tips contrast with the gray of the rest of the pod.

• When arranging, keep in mind that these pods appear more decorative from above than in profile; however, both views are appealing.

Deptford Pink (Weber)

FLOWERS

Bloom: June–Sept. 1–2½ ft. tall
Flowers: Orange-yellow, in clusters,
         edges sometimes black-spotted
Leaves: Linear-oblong, opposite
Found: Disturbed areas, roadsides,
        neglected fields

ST. JOHNSWORT
Hypericum perforatum

PODS (Aug. to frost)

Let these pods drop their leaves and dry in the field before they are picked. Weather conditions and different areas will determine colors that vary from rich auburn reds to muted browns.

• The dense cluster shape of these small, pointed pods provides good filler in bouquets; also, be aware of the interesting, clean, fine stem formation for use in large, open sprays.

St. Johnswort, Multiflora Rose hips, Poor-man's Pepper (Weber)

DAISY FLEABANE
Erigeron annuus

FLOWERS

Bloom: June–Sept. 1–3 ft. tall
Flowers: Fine white rays, yellow disks, in branching clusters
Leaves: Sharply toothed, lower leaves ovate, upper leaves narrow
Found: Fields, roadsides, disturbed areas; very common

PODS (Aug. on)

Fleabane with its very small pods appears airy and fragile. For curry-yellow color pick flowers and hang to dry. If left in the field too long, the plant becomes brittle, and upper branches easily break off.

• It is an excellent little species to use in miniatures or as a filler in larger arrangements.

Daisy Fleabane, artificial pears, dried fern leaves (Weber)

FLOWERS

Bloom: July–Oct. 6–24 in. tall, bushy
Flowers: Greenish, numerous, on nearly
leafless upper branches
Leaves: Long, serrated, fall early
Found: Cultivated fields, roadsides,
sandy soils

WINGED PIGWEED
Cycloloma atriplicifolia

PODS (Aug.–Nov.)

This amazing tumbleweed-type bush turns all sorts of lovely colors as it rolls across the countryside—greens, wheat, deep straw colors, and unbelievable shades of red.

• Use a total, shapely bush in attractive pottery for a conversation piece; it is also good used as a filler when broken into smaller pieces.

Entire Winged Pigweed plant

65

FLOWERS
Bloom: July–Oct. 2–5 ft. tall
Flowers: Greenish, small, nodding, long
      slender clusters
Leaves: Forked, stringy, fine, long
Found: Sandy soil, especially along
      eastern coast

TALL WORMWOOD
Artemisia caudata

PODS (Aug.–Oct.)
These lovely, tapered, flowering heads form the shape of the entire plant. Picked anytime, they dry beautifully. The showy colors range from pale and amber greens to honey browns and nut browns.

• With this versatile plant the creative possibilities are many. Take advantage of its variety in color and shape; use horizontally or vertically and include in large or small arrangements.

Tall Wormwood, mixed nuts
(Weiler)

66

FLOWERS

Bloom: July–Oct. 2–6 ft. tall
Flowers: Lavender, minute florets on egg-
         shaped heads
Leaves: Paired, lanceolate, embrace
         prickly stems
Found: Roadsides, disturbed areas, old
         pastures; common

TEASEL
Dipsacus sylvestris

PODS (Aug. on)

Here is a sturdy pod that can be picked all winter. It looks like a honeycombed egg nestled within a long, spiny bract nest. Early pods are a light honey color; the longer the pods remain in the field, the darker brown they become.

• Although extremely prickly to handle, Teasel has a pleasing shape to use in arrangements, wreaths, wall hangings, and ornaments.

Teasel, Honey Locust pods

WILD YARROW
Achillea millefolium

FLOWERS
Bloom: June–Aug. 1–3 ft. tall
Flowers: White, flat, dense clusters
Leaves: Fernlike, finely divided, soft,
aromatic
Found: Disturbed areas, roadsides,
fields

PODS (mid-Aug.–Oct.)
Picked early, the pods are a warm golden
brown; later they turn a deeper brown.
Yarrow has several neat, tight clusters
that form the main pod head. (These dried
pod heads are often used by architects
as trees in model presentations.)
• It is an attractive cluster for mixed arrangements and is easily separated for
smaller bouquets.

Sprayed Wild Yarrow, Baby's Breath

FLOWERS
Bloom: June–Aug. 2–4 ft. tall
Flowers: Pink to lavender, rosettes in leaf
   axils
Leaves: Deeply cut into 3 points,
   opposite and nearly horizontal
Found: Disturbed areas, roadsides;
   common

MOTHERWORT
Leonurus cardiaca

PODS (mid-Aug.–Oct.)
An attractive pod to look at with its spiny,
burlike clusters that graduate in size on
the square stem. It is difficult to handle
because of its prickly quality. The loca-
tion and weather determine the varying
shades of brown and gray.
• This pod provides excellent textured
verticals in arrangements despite a few
thorns in handling.

Motherwort, Great Ragweed, Queen
Anne's Lace, Plume grass, miscellaneous
grasses (Weiler)

FLOWERS

Bloom: June–Sept. 1½–4 ft. tall
Flowers: Bright yellow rays always
        drooping down, spotted,
        elongated button centers
Leaves: Deeply lobed, hairy, on hairy
        stems
Found: Prairies, fields, open woods

GRAY-HEADED CONEFLOWER
Ratibida pinnata

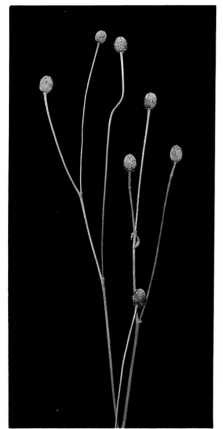

PODS (mid-Aug.–Oct.)

Gray-headed Coneflowers are conspicuous by their gray oval heads spotted by
dark stubble. As the pods age, they appear pitted. They are at their peak color if
picked as soon as seeds drop.

• Remember this pod if deep gray tones
are needed for an arrangement.

Gray-headed Coneflower, Ilex
Holly (Weber)

FLOWERS

Bloom: July–Sept. 1–3 ft. tall
Flowers: Mustard yellow, button-shaped, flat-topped clusters
Leaves: Fernlike, aromatic, ornamentally toothed
Found: Roadsides, disturbed areas, open fields; common

COMMON TANSY
Tanacetum vulgare

PODS (mid-Aug.–Oct.)

Tansy may be picked in the flower stage, and it will dry to all shades of mustard yellow. In the field it develops velvety, flat-button pods ranging in color from tawny to chocolate browns.

• These wonderful, flat, compact button clusters have pleasing shapes and add strong yellows and browns to any arrangement.

Common Tansy, grass leaves

FLOWERS

Bloom: June–Sept. 1–5 ft. tall

Flowers: Soft yellow, 4 petals, flowering up very long spikes

Leaves: Pale green, lanceolate, hairy, velvety-looking

Found: Disturbed areas, dry sandy soil

COMMON EVENING-PRIMROSE
Oenothera biennis

PODS (mid-Aug. on)

In the field it is easy to spot Common Evening-primrose by the tall spikes that stand above all other flowers. Tubular pods 1 in. long with 4 tips climb alternately up the stem, maturing from warm light-brown to gray-brown shades.

• This pod may be used in a vertical or horizontal manner and works well in sturdy floor arrangements.

Common Evening-primrose, Fireweed, Oats, grasses, pine cones (Weber)

FLOWERS

Bloom: Apr.–Aug. 1–1½ ft. tall
Flowers: White, with inflated pinkish calyx sacs
Leaves: Opposite, pointed oval
Found: Disturbed areas, roadsides, fields; common

BLADDER CAMPION
Silene cucubalus

PODS (late Aug.–Sept.)

Pods look like small bleached tulip buds, with *5 teeth not curled* on top. They are usually V-branched with a single pod in the crotch.

• Bladder Campion is very light-colored and has a delicate quality when used alone, mixed, or in miniature arrangements.

Bladder Campion, Bull Thistle, grasses (Weber)

LEADPLANT
Amorpha canescens

FLOWERS
Bloom: June–Aug. 1–3 ft. tall
Flowers: Purple, tipped in orange, dense spikes
Leaves: Pinnate, small, numerous, hairy gray appearance.
Found: Open woods, prairies

PODS (end of Aug.–Sept.)
The pod spikes look purplish gray and, if picked as soon as pods form and hung to dry, spikes will retain their fullness. If too matured, spikes tend to lose their pods, and interestingly colored, spotty-textured, sticklike spikes remain.

• The purple-gray cast adds an unusual color to fall dried materials and lends itself well to use in pewter and silver containers.

Leadplant, Purple Prairie Clover, Nodding Thistle (Weiler)

74

FLOWERS
Bloom: June–July 1–3 ft. tall
Flowers: Creamy yellow, very upright
Leaves: 7 to 11 leaflets on hairy stems
Found: Rocky soil fields, dry woods

TALL CINQUEFOIL
Potentilla arguta

PODS (end of Aug.–Sept.)
These pods have a very close-clustered head, giving them a compact, wood-sculptured look. If picked early, pods are a soft yellow-brown; later they become a more drab brown.
• Because of its compactness, this pod works well in wreaths or in short or tall arrangements. It has great versatility.

Tall Cinquefoil, Baby's Breath (Weiler)

FLOWERS

Bloom: June–July 1–3 ft. tall
Flowers: Deep pink, 5 large petals
Leaves: Compound, 7 oval leaflets,
prickly stems, straight thorns
Found: Sandy soil, roadsides, open
woods

PASTURE ROSE
Rosa carolina

PODS (end of Aug.–Oct.)
Early Pasture Rose hips are orange to cherry red; later they become scarlet to ruby red. Whenever they are picked, keep cool and dry while drying, as rose hips tend to spoil if conditions are too damp. The dried, good-size, shriveled berries with black tips are beautiful in texture and shape.
• This is an excellent, showy red accent for the usual muted colors of dried pods.

Pasture Rose hips, Baby's Breath

FLOWERS

Bloom: July–Sept. 2–4 ft. tall
Flowers: Pink to lilac, shaggy-looking
Leaves: Lanceolate, opposite, toothed
Found: Open woods, dry fields,
roadsides

WILD BERGAMOT
Monarda fistulosa

PODS (end of Aug.–Oct.)

These lovely pods have dusty brown, honeycombed, round heads. Pick early for brighter tones. If picked late, pods are colorless but can be sprayed with stain to highlight the color.

• Bergamot pods add nice ball-shaped accents and add solidity to arrangements.

Wild Bergamot, Wild Lettuce,
Mountain-mint (Weber)

77

FLOWERS

Bloom: July–Oct. 6–30 in. tall
Flowers: Pink to red, fuzzy
Leaves: Linear, alternate
Found: Dry prairies, open areas, sandy
       soil

BLAZING STAR
Liatris aspera

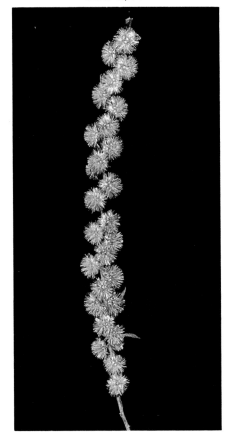

PODS (late Aug.–Oct.)

These plants have blunt spikes with fuzzy pods coiling upward. If possible, spray before picking and carefully hold branches apart. Pull off seed fuzz on some stems to obtain a strawflower effect for variety.

• These straight spikes entwined with furry pods used as the center of interest are excellent for vertical arrangements.

Blazing Star, Bur-reed, dyed
Starflowers

FLOWERS

Bloom: June–Oct. 1–5 ft. tall
Flowers: Yellow, dandelionlike, on
  numerous branches near top
Leaves: Oblong, pointed lobes where
  they embrace stems
Found: Disturbed areas, roadsides,
  fields; common

COMMON SOW-THISTLE
Sonchus oleraceus

PODS (late Aug. on)

Pick Sow-thistle as soon as all blooming
has stopped but some white, cottony seed
heads still remain. Spray and hang to dry.
• The dark, irregular pods with some
touches of white on the crooked, dark
stems add a wavy, casual air to large or
small arrangements.

Common Sow-thistle, garden Yarrow, German Statice, Strawflowers

79

SUN

COMMON BURDOCK
Arctium minus

FLOWERS
Bloom: July–Oct. 1½–5 ft. tall
Flowers: Pink-lavender florets atop thistlelike burs
Leaves: Lower leaves large, heart-shaped
Found: Disturbed land, roadsides; very common

PODS (late Aug. on)
These bur heads are easily spotted on sturdy, almost bush-size plants. Handle very carefully and avoid getting on clothes. Keep stems separated, as burs stick together when they touch each other.
• Since Burdock with its rough texture and rich browns makes good contemporary arrangements, the extra care in handling is worthwhile.

Common Burdock, Corkscrew Willow, dried leaves (Weber)

80

FLOWERS
Bloom: June–Oct. 1–4 ft. tall
Flowers: Ice blue, sometimes white,
square-tipped, close by noon
Leaves: Basal leaves dandelionlike
Found: Roadsides, disturbed land,
pastures; very common

CHICORY
Cichorium intybus

PODS (end of Aug. into winter)
Some pods and branches maintain a lovely silhouette; others look straggly. Some areas produce warm brown colors; other areas, drab gray-browns. The later the pods are picked, the grayer the tone.
• These curved, podded branches look graceful used alone or included in a mixed bouquet.

Chicory

FLOWERS

Bloom: June–Sept. 2–7 ft. tall
Flowers: Purple-pink on tall racemes
Leaves: Alternate, lanceolate
Found: Flourish in newly burned areas
(hence name), roadsides, low
meadows

FIREWEED
Epilobium angustifolium

PODS (Sept.–Oct.)

The pods are unusual curly-looking spikes.
In both color and texture they look like
rods of excelsior. If picked early, some still
may have silky white seed threads entwined. If pods are left until later, they are
whipped clean by the winds.
• This long-stemmed pod adds a unique
shape and texture to any bouquet.

Fireweed, sprayed Yellow Goat's-beard
skeletons, dried Daisy, garden Love-in-
a-mist

FLOWERS

Bloom: June–Sept. 1–2 ft. tall
Flowers: White to pink, 5 petals,
transparent green calyx sacs
Leaves: Opposite, long, hairy
Found: Roadsides, disturbed areas;
common

WHITE CAMPION
Lychnis alba

PODS (Sept.–Oct.)

Pods are ready to pick when leaves dry and pods have shed their skin. They are cuplike, tipped with *10 curled teeth,* nut brown and waxy-looking, perched on branched stems. If left in the field too long, the plant tends to sprawl.

• This shiny, sturdy pod adds snap to dried materials and is easily arranged.

White Campion, pheasant feathers

83

CARRION-FLOWER
Smilax herbacea

FLOWERS
Bloom: May–June Vines to 7 ft. long
Flowers: Yellow-green, clustered, foul
odor
Leaves: Heart-shaped, parallel veins
Found: Fence lines, woods, over bushes

PODS (Sept.–Oct.)
The vines have beautiful balls of solidly packed blue-black berries. When picking *any* berry pods, it is better to wait until leaves are dry, because the berry is less likely to rot when brought indoors.
● For use in a striking oriental or asymmetrical arrangement, use some clusters still attached to the vine with its tendrils.

Carrion-flower (Weber)

FLOWERS
Bloom: May–June 1–4 ft. tall
Flowers: Yellow, tiny, bell-shaped
Leaves: Wispy, fine
Found: Along fence lines, roadsides

WILD ASPARAGUS
Asparagus officinalis

PODS (Sept.–Oct.)
Some plants have burnt-orange berries. Most Asparagus leaves become showy in marvelous colors of green, straw yellow, and pumpkin orange. Pick when they are the desired color. Hang to dry and to keep the shape. Spray leaves to check shedding.
• The fernlike quality of this plant is good for a filler or for edging bouquets.

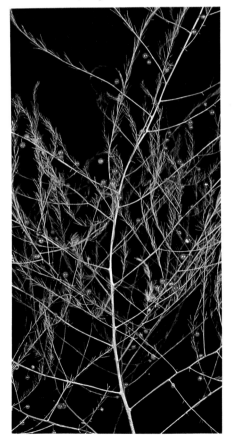

Wild Asparagus with
Berries

FLOWERS
Bloom: June–Sept. 1–3 ft. tall
Flowers: Pale pink-purple, scalelike
bracts similar to Bachelor
Buttons
Leaves: Lanceolate, on very branching
stems
Found: Disturbed areas, dry fields,
roadsides, in patches; common

SPOTTED KNAPWEED
Centaurea maculosa

PODS (Sept.–Oct.)
These plants spread easily and are usually
found in masses. There are several species
worth mentioning whose pods differ in
color, traits, and shapes. Spotted Knapweed (see pod photo) has dark, fringed
tips on bracts; Tyrol Knapweed, *C. vochinensis,* has paler, more pointed pods;
Brown Knapweed, *C. jacea,* has darker,
stubby pods with rounded scales on
bracts.
• All Knapweeds are decorative, interesting pods.

Spotted Knapweed, dyed imported Centaurea

FLOWERS

Bloom: June–Sept. 2–4 ft. tall, bushy
Flowers: Creamy white, pealike, on spikes
Leaves: Pinnate, alternate, smooth soft green
Found: Dry sandy soil, clearings; common in Great Lakes area

EARLY FALSE INDIGO
Baptisia leucophaea

PODS (early Sept.–Oct.)

Plants are usually erect and bushy and easy to identify by the pecan-size, smooth, hard-shell blue pods whose seeds rattle inside. Wild Indigo, *B. tinctoria,* another often found species, has yellow flowers in looser racemes, and its leaves turn black when dried.

• These extremely unusual pods add special interest in color and shape to any arrangement.

Early False Indigo

LAMB'S-QUARTERS
Chenopodium album

FLOWERS

Bloom: June–Oct. 1–4 ft. tall

Flowers: Green, ball-like, at ends of branches and axils of leaves

Leaves: Diamond-shaped, silvery sheen, variable

Found: Roadsides, cultivated fields, disturbed areas; very common

PODS (Sept.–Oct.)

In the fall the tiny seed balls turn from green to extraordinary shades of red and cover the roadsides with their vivid colors. Later they turn to varying shades of gray.

● This nondescript plant which most bypass as "just a weed" has great color, a nicely branched spikelike form, and a mini-popcorn-seed texture for use in arrangements.

Lamb's-quarters, dried green Gum-weed, Purple Prairie Clover, Love grass (Weber)

FLOWERS
Bloom: June–Oct. 2–7 ft. tall
Flowers: Magenta-rose, large drooping
        clusters
Leaves: Heart-shaped, large, long-
        stalked leaf stems
Found: Disturbed areas, open areas

PRINCE'S-FEATHER
Polygonum orientale

PODS (Sept.–Oct.)
This plant has surprisingly vivid magenta
blooms of majestic size. As the blooms
begin to pod and to capture some of the
rosy color, pick and hang to dry. The stems
and leaves dry to colorful reds that are
most desirable.
● There is a subtle arching, drooping, oriental quality to these pods which suits
them well to contemplative arrangements.

Prince's feather

89

FLOWERS
Bloom: June–Oct. 2–3 ft. tall
Flowers: Red-purple, large, daisylike,
darker-colored spiny centers
Leaves: Long, stalked, rough-toothed
Found: Dry open woods, prairies, dry
fields

PURPLE CONEFLOWER
Echinacea purpurea

PODS (Sept.–Oct.)
Purple Coneflowers are most attractive as soon as rays wither, because pod centers are then a rich burgundy-brown to black in color. Pods are large; some are almost walnut-size on sturdy stems, and heads have a porcupinelike texture.
• Because of its size and dark, spiny quality, this is an impressive pod to use in mixed arrangements.

FLOWERS

Bloom: July–Aug. 1–2 ft. tall
Flowers: Lavender, dome-shaped umbels, crooks at top of stems
Leaves: Grasslike, upright from base
Found: Roadsides, disturbed areas, fields

NODDING WILD ONION
Allium cernuum

PODS (early Sept.–Oct.)

Pick some Nodding Wild Onion while still partially in bloom, and hang to dry. Thus treated, the heads seem to retain a better shape. The little black ball seeds are cradled in the opened pods and give the pod an airy, salt-and-pepper, pompon appearance.

• This pod is unusual, round yet open in shape, giving a perky accent to bouquets.

Nodding Wild Onion, Yarrow, Yellow Rocket, Velvet-leaf, Millet (Weiler)

FLOWERS

Bloom: July–Sept. 1–2 ft. tall

Flowers: Bright orange, numerous, in umbels

Leaves: Oblong, alternate, on hairy stout stems

Found: Dry soil, open fields, railroad embankments

BUTTERFLY-WEED
Asclepias tuberosa

PODS (Sept.–Oct.)

As for most Milkweeds, for better color and for flamboyant contours it is better to allow the plants to mature and release their seeds in the field. If picked green and dried indoors, their forms are more stilted and restrained.

• Butterfly-weed pods burst open into shapes that resemble burnished orchids coated with soft velour exteriors. It is an exquisite pod for arrangements.

Butterfly-weed, Yellow Rocket

92

FLOWERS
Bloom: July–Sept. 4–8 ft. tall
Flowers: Bright yellow rays and disks
Leaves: Large, in pairs, surrounding
        stems to form cups
Found: Fencerows, roadsides, disturbed
       areas

CUP-PLANT
Silphium perfoliatum

PODS (Sept.–Oct.)
This pod is particularly beautiful picked
green and dried. It retains its lovely color
and looks like a green opened rose. The
dried silver-brown pods are interesting
when some seeds are gone and smooth,
shallow-cup flowers remain.
• This large, long-stemmed pod in soft
greens and browns is charming in colonial
bouquets.

Cup-plant, Baby's Breath, Statice, Multiflora
Rose hips (Weiler)

93

COMPASS-PLANT
Silphium laciniatum

FLOWERS

Bloom: July–Sept. 4–9 ft. tall
Flowers: Yellow rays, yellow disks, bristly
        bracts
Leaves: Huge, 1–3 ft., deeply cut to ribs,
        stiff, hairy
Found: Dry and wet prairies, open areas

PODS (Sept.–Oct.)
This unusually striking plant is particularly
distinctive for its huge, rough, coarsely cut
leaves that develop beautiful, stiff, con-
torted shapes when dry. Let nature take its
course to shape the contours! The pod is
much like other *Silphiums,* such as Cup-
plant and Prairie-dock, but much larger on
a bristly stem.

- These artistic leaves and pods are excellent for arrangements in shallow containers.

Compass-plant leaves, Yellow Goat's-
beard (Weiler)

FLOWERS

Bloom: July–Sept. 3–8 ft. tall
Flowers: Pale yellow, tiny and
  dandelionlike, in elongated
  panicles
Leaves: Deeply lobed, lance-shaped
Found: Open areas, roadsides, thickets;
  common

WILD LETTUCE
Lactuca canadensis

PODS (Sept.–Oct.)

The older this plant becomes, the more interesting is its shape and texture. All branches turn upward, giving a vertical, linear appearance. Picked early, Wild Lettuce has smooth, light-cork-colored stems sprinkled with tiny gray-white pods. It can be picked throughout the winter.

• Unusual effects may be achieved with this shapely wildflower.

Wild Lettuce

95

FLOWERS
Bloom: July–Sept. 2–6 ft. tall
Flowers: Reddish purple, flat clusters
Leaves: Lanceolate, opposite, sessile, numerous
Found: Pastures, prairies, disturbed areas

WESTERN IRONWEED
Vernonia fasciculata

PODS (Sept.–Oct.)
This stately, coarse plant generally gets very large. The flat-topped, open cluster may be 8–10 in. across, with each individual pod appearing as a stiff, miniclover head.

● These nicely proportioned plants, with spiny, textured pods and honey-brown color, make a sizable species excellent for use in outsized or floor arrangements.

Western Ironweed, Silver Dollars, Fox-tail grass

FLOWERS

Bloom: July–Sept. 1–3 ft. tall, bushy
Flowers: Dull yellow with purple-brown throats
Leaves: Diamond-shaped, smooth
Found: Open woods, disturbed areas, meadows

SMOOTH GROUND-CHERRY
Physalis subglabrata

PODS (Sept.–Oct.)

Smooth Ground-cherry pods cling much better to the branches if picked as pale green Chinese lanterns and dried in normal upright position. If allowed to mature in the field to wheat color, pods readily drop. Handle with care as they tangle easily.

• With or without stems, this pod has endless decorating possibilities.

Smooth Ground-cherry
(Weiler)

97

FLOWERS
Bloom: Mid-July–Oct. 2–6 ft. tall
Flowers: Deep yellow, yellow disks, daisylike
Leaves: Short, arrowhead-shaped, thick, rough, hairy stems
Found: Open woods, roadsides, dry soil, thickets

STIFF-HAIRED SUNFLOWER
Helianthus hirsutus

PODS (Sept.–Oct.)
In the late summer these tall, deep yellow flowers with coarse, scratchy leaves are common along the country roadsides. To obtain lovely nutmeg color and clean, spiny, globular heads, pick as soon as they pod.

• Stiff-haired Sunflower is a rough, tough, but nicely shaped pod with an appealing color that adds warmth to arrangements.

Stiff-haired Sunflower
(Weiler)

FLOWERS
Bloom: July–Sept. 4–10 ft. tall
Flowers: Orange-yellow, on very long
reddish stems
Leaves: Huge, up to 2 ft., oval, heavy-
looking, at bases of plants
Found: Prairies, meadows, open areas

PRAIRIE-DOCK
Silphium terebinthinaceum

PODS (Sept.–Oct.)
Begin early to look for the leaves that turn
leather brown and become beautifully
curled and twisted shapes of stiff parchment. The pods may still be green at that
time, but some can be picked and dried for
color. Later, collect the brown, open-rose-
shaped pods.
• The dramatic leaves and the long-
stemmed flower pods have endless creative possibilities.

Prairie-dock

FALSE BONESET
Kuhnia eupatorioides

FLOWERS
Bloom: Aug.–Sept. 1–3 ft. tall
Flowers: Off-white, rayless
Leaves: Lanceolate, variable, alternate
Found: Dry fields, open woods

PODS (Sept.–mid-Oct.)
For a better seed head that is less likely to shed, this pod should be picked in bloom and hung to dry. Field-matured pods also are lovely, however, and the fuzzy gray-white color is a distinctive characteristic. Spray before using.

● This soft white pod is particularly attractive when used in Williamsburg bouquets.

False Boneset, Curled Dock, Teasel, garden Yarrow, Turtlehead, dried leaves (Weber)

FLOWERS

Bloom: Aug.–Sept. 2–4 ft. tall
Flowers: Golden yellow, flat-topped, loose
multiple clusters
Leaves: Willowlike, slender, very small
Found: Damp areas, riverbanks, thickets

LANCE-LEAVED GOLDENROD
Solidago graminifolia

PODS (Sept.–Oct.)

These Goldenrods grow in masses in wet areas. The loose, flat clusters of yellow heads turn a light toasty brown, with many smaller clusters making up a large central head. The flowers dry a beautiful yellow.

• This species is smaller and not as dense as other Goldenrods and adapts well to any size arrangement.

Lance-leaved Goldenrod (Weiler)

101

FLOWERS

Bloom: Aug.–Oct. 2–5 ft. tall
Flowers: Yellow, clusters at ends of elmlike arching branches
Leaves: Elliptical, coarsely toothed, alternate
Found: Dry woods, thickets, wooded roadsides

ELM-LEAVED GOLDENROD
Solidago ulmifolia

PODS (Sept.–Oct.)

Elm-leaved Goldenrod has long, arching branches with flower clusters resembling an exaggerated yellow caterpillar sunning on the end of each branch. Pick them in bloom or when matured, and hang to dry.

• The soft tan of the fuzzy dried heads and the pleasing shapes of this particular Goldenrod dress up any bouquet.

Elm-leaved Goldenrod, Curled Dock, False Boneset, grasses (Weber)

FLOWERS
Bloom: Aug.–Oct. 1–5 ft. tall
Flowers: Green, in dense spikes in upper
        leaf axils, chaffy
Leaves: Ovate, dull green, on long stalks
Found: Disturbed areas, cultivated
        fields, roadsides

GREEN AMARANTH
Amaranthus retroflexus

PODS (Sept.–Oct.)
Most people overlook the creative possibilities of this rough species (also known as Pigweed). It should be considered as a decorative plant because of the lovely colors it retains and because of the variability of its growth pattern.
• Just a pot of Amaranth's variations of green and brown makes an interesting, robust arrangement.

Green Amaranth, Corkscrew Willow,
pine cones

SUMMER CYPRESS
Kochia scoparia

FLOWERS
Bloom: Aug.–Oct. Up to 4 ft. tall
Flowers: Small, light green, small auxiliary clusters
Leaves: Lance-shaped, narrow, red in maturity, pyramidal branched plants
Found: Roadsides, sandy soil, disturbed areas

PODS (Sept.–Oct.)
Summer Cypress is another overlooked fall wildflower; yet its airy pyramidal shape turns the roadside into swaying foliage of brilliant colors. It is not only the pod but the entire colorful plant that is decorative; the stems particularly in crimson, scarlet, and cardinal reds do much to enhance arrangements.
• This pod is so plentiful; use it abundantly in masses or as spike accents in bouquets.

Summer Cypress, Common Milkweed, grasses (Weber)

FLOWERS

Bloom: Aug.–Oct. 1–2 ft. tall
Flowers: Small, off-white, clustered, fragrant
Leaves: Long, linear, alternate, cottony underneath
Found: Dry soil, old fields, pastures

SWEET EVERLASTING
Gnaphalium obtusifolium

PODS (early Sept. into Oct.)
Flowers may be picked and hung to mature. Pods are more straw-colored than the white Pearly Everlasting and do not have as dark a center. Pods are aromatic even when dry. Shake off seeds for strawflower effect and for a deeper color.
• This is another beautiful pale pod to add to fall arrangements.

Sweet Everlasting, Little Bluestem grass

BEGGAR-TICKS
Bidens frondosa

FLOWERS

Bloom: Aug.–Oct. 1–4 ft. tall
Flowers: Sulfur yellow, heads have no rays
Leaves: 3–5 leaflets, opposite, toothed
Found: Fields, damp soil, ditches; very common

PODS (Sept.–Oct.)

Pick some flowers for yellow-green color and hang to dry. Later the pods develop an interesting barbed look. Handle carefully; seeds have 2 barbed awns and readily stick to anything. Swamp Beggar-ticks, *Bidens connata,* have lanceolate leaves and are found in wet areas.

• Spray to ease arranging problems. Combine dried flowers and pods in arrangements.

Beggar-ticks and Swamp Beggar-ticks, sliced pine cone flowers, dyed Starflowers

106

FLOWERS

Bloom: June–Sept. 2–6 ft. tall
Flowers: Pale yellow, 5 petals, on clublike stalks
Leaves: Large, oblong, wooly
Found: Poor soil, roadsides, disturbed areas; common

COMMON MULLEIN
Verbascum thapsus

PODS (Sept.–Nov.)

The amber-brown, tall, columnlike stalks are unique and easy to spot in the fields. Pick some while still pale green and some very small stalks for smaller arrangements.

• There is a velvety, textural quality to Mullein. It adds very strong vertical accents and is interesting used in massive floor arrangements or in heavy pottery.

Dried green Common Mullein,
Tall Upland Boneset, grapes
(Weber)

107

VELVET-LEAF
Abutilon theophrasti

FLOWERS
Bloom: July–Oct. 2–5 ft. tall
Flowers: Pale pumpkin yellow, 5 petals on short stalks in leaf axils
Leaves: Heart-shaped, velvety, very large
Found: Edges of soybean and corn fields, disturbed areas

PODS (Sept.–Nov.)
The pods are cup-shaped with prickles on upper edge and look like butter print blocks. Pick when leaves are dry and pods are dark. If dusty from fields, spray with water to regain deep charcoal color. Pods are on light brown, velvety stems.

• This decorative, almost black pod is an impressive accent in arrangements or is dramatic when used alone.

Velvet-leaf

FLOWERS

Bloom: Aug.–Oct. 2–6 ft. tall
Flowers: White, numerous rays, yellow-
orange disks
Leaves: Long, slender, alternate
Found: Marshes, damp open areas

MARSH ASTER
Aster simplex

PODS (Sept.–Nov.)

This species is an especially nice Aster pod
with its fuzzy tops in little round clusters at
the tips of many branches.

● Its lovely light color brightens the more
usual monotone of dried arrangements.
The growing pattern of pod heads shooting off the main stem is in itself decorative.

Marsh Aster

FLOWERS

  Bloom: July–Nov. 1–6 ft. tall
Flowers: Greenish white on numerous
         stalks near top
 Leaves: Linear, dark green, very leafy
         stems
  Found: Roadsides, disturbed areas;
         common

HORSEWEED
Erigeron canadensis

PODS (Sept.–Nov.)

Once the adjacent photos have been stud-
ied, Horseweed will be easily identified in
fields. The mature arrowhead-shaped top
is loaded with numerous, fuzzy, seeded
heads and white, pinhead-size, weedless
pods. Pick some green plants and hang to
dry for color variation.

• These airy, textured heads may be used
creatively alone or broken into smaller
clusters for mixed arrangements.

Horseweed, Boneset

FLOWERS

Bloom: July–Sept. 2–6 ft. tall
Flowers: Pale yellow, long rays, yellow disks
Leaves: Paired, rough, sessile on ruddy stems
Found: Dry or sandy soil, roadsides, prairies

ROSINWEED
Silphium integrifolium

PODS (early Sept.–Nov.)

This large, coarse plant has the usual qualities of *Silphium* pods; it is lovely when dried green. If it is dried in the field, the seeds drop and a rough cluster of silver-brown flowers with pointed petals and small button centers remain.

• Used as a pale green cluster or as a soft brown open-petal flower, Rosinweed adds color and textural assortment to dried materials.

Rosinweed, Strawflowers, Willow Herb, Round-headed Bush-clover, dried Oak leaves

111

FLOWERS
Bloom: June–Aug. 1–2 ft. tall
Flowers: Pale yellow, in flat clusters
Leaves: 5–7 radial leaflets coarsely
   toothed
Found: Roadsides, fields, meadows;
   common

ROUGH-FRUITED CINQUEFOIL
Potentilla recta

PODS (Sept. into winter)
These pods look like tiny rosebuds
branched in a somewhat flat-headed, airy
cluster. The color is usually a gray-tan. Let
mature in the field but pick early for better
color and form.
• For a nice crisp look use this pod alone
or as a filler in mixed arrangements.

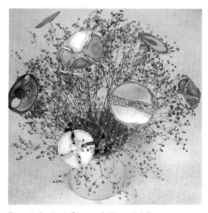

Rough-fruited Cinquefoil, metal flowers

FLOWERS

Bloom: July–Oct. 1–3 ft. tall
Flowers: Pale yellow rosettes in upper leaf axils, cupped by showy purple bracts
Leaves: Lanceolate, opposite, slightly toothed
Found: Sandy soil, open areas; common along coastal area

HORSEMINT
Monarda punctata

PODS (Sept.–Nov.)
Horsemint in bloom has an almost exotic appearance with its layered, pale green and purple bracts. Like other *Monardas,* it develops nearly black, round pod heads which appear to be pierced by the stem.

• The globular pods climbing upward on branching stems are distinctive in any bouquet.

Horsemint, Statice, Goldenrod

113

HARD-LEAVED GOLDENROD
Solidago rigida

FLOWERS

Bloom: Aug.–Oct. 2–5 ft. tall
Flowers: Deep yellow, flat-topped, dense clusters
Leaves: Oval, extremely rigid, upper clasping stems pointed upward, hairy
Found: Dry prairies, sandy soil, open areas

PODS (Sept.–Nov.)

Hard-leaved Goldenrod grows in clumps and stands out with its full, dense heads and hard, upper oval leaves clutching the stems. Picked and dried late in bloom, it pods in beautiful yellows; picked when mature, the heads are lovely, fuzzy, flecked gray-green beige.
• The fullness of these heads haughtily balanced on very stiff stems adds character to any bouquet.

Hard-leaved Goldenrod, Tumble Mustard, Chicory, Tall Cinquefoil

FLOWERS

Bloom: July–Sept. 1–15 ft. tall
Flowers: Green, tiny, dense racemes
Leaves: In pairs, 3 sharp-pointed lobes
Found: Roadsides, dry areas; common

GREAT RAGWEED
Ambrosia trifida

PODS (Sept.–Nov.)

As flowers these huge roadside plants are not impressive; as pods they present an unusual shape. Great Ragweed at the top of its stalks terminates in 5 or more shorter stems, each adorned with clusters of dried, ruffly pods that seem pierced by bearded spikes.

• This tawny brown pod is an exciting shape in arrangements and is abundantly available everywhere.

Great Ragweed, Needle
grass (Weiler)

115

FLOWERS

Bloom: July–Sept. 1½–4 ft. tall
Flowers: White, tiny, on round heads
Leaves: Stiff, upright, elongated-lanceo-
late, spiny edges
Found: Dry or moist open woods,
thickets, prairies

RATTLESNAKE-MASTER
Eryngium yuccifolium

PODS (Sept.–Nov.)

This unusual plant is identified by the yucca or cactuslike appearance of its base leaves. The tall, grooved stems hold an open cluster of round, pitted pods about the size of acorns. These pods keep well and are sturdy as well as decorative.

• The lovely sable brown of this pod with its rough texture will dress up any arrangement.

Rattlesnake-master, Prairie Cord grass

FLOWERS

Bloom: July–Sept. 1–2 ft. tall
Flowers: Greenish white, in clusters near top of stems
Leaves: Linear, light green, numerous
Found: Fields, roadsides, sandy soil

WHORLED MILKWEED
Asclepias verticillata

PODS (mid-Sept.–Oct.)

Pick when stems are brown, the long, tapered pods have burst open, and seeds are airborne on feathery parachutes. The remaining pods look like parchment butterflies perched on long, delicate stems. If left in fields too long, pods become mottled and drab in color.

• Whorled Milkweed pods add a light, sparkling touch to bouquets.

Whorled Milkweed, Goldenrod insect galls, Frost Aster

117

FLOWERS

Bloom: June–Aug. 1–4 ft. tall

Flowers: Pale pink, bell-shaped, deep pink stripes inside

Leaves: Ovate, opposite, on red-tinted stems

Found: Roadsides, thickets, edges of woods

SPREADING DOGBANE
Apocynum androsaemifolium

PODS (mid-Sept.–Oct.)

This plant is shrublike. Seedpods hang in pairs like wishbones at the end of curved branches. Pods are approximately 2 in. long and very tapered. Wait to pick until pods are reddish brown.

• They are lovely used alone in a container, or they add graceful lines to simple arrangements.

Spreading Dogbane

FLOWERS

Bloom: July–Sept. 1–3 ft. tall
Flowers: Blue or violet, 5-pointed, bells
hanging along *one* side of stems
Leaves: Heart-shaped lower leaves,
upper ones oval, pointed
Found: In patches, roadsides, fields,
especially New England area

CREEPING BELLFLOWER
Campanula rapunculoides

PODS (mid-Sept.–Oct.)

Pods are ready to pick when bell-shaped
and stems are brown. Parchmentlike pods
hang straight down one side of the stiff
stem. If picked late, pods become dirty-
looking and unattractive.

• The erect, slender stems add dainty
shoots to any bouquet.

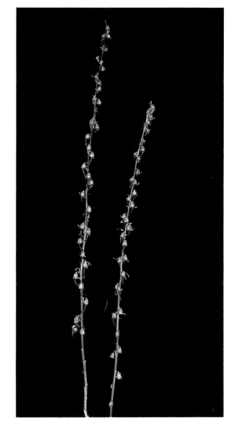

Creeping Bellflower, Round-headed
Bush-clover, Tansy, Hoary Alyssum
(Weiler)

119

FLOWERS

Bloom: Early Sept.–Oct. 1–3 ft. tall
Flowers: Fine white rays, yellow centers
Leaves: Narrow, small, numerous, on plume-shaped branches
Found: Dry sandy soil, open meadows, roadsides

FROST ASTER
Aster pilosus

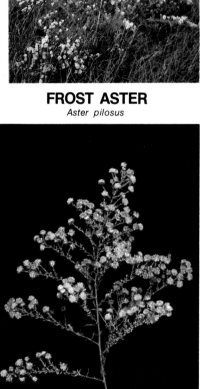

PODS (mid-Sept. to frost)
With the tufts of silky seeds attached and sprayed to prevent shedding, pods are fuzzy and interesting to use; or shake to remove seedpods, and starlike strawflowers will remain. This is true of most Asters.

● Both types of pods should be considered for arrangements and can be used like Baby's Breath.

Frost Aster

FLOWERS
Bloom: June–Aug. 2–5 ft. tall
Flowers: Rose-pink, ball-shaped clusters, fragrant
Leaves: Broadly oval, rounded tips, opposite, thick
Found: Roadsides, dry fields, gravelly soil; common

COMMON MILKWEED
Asclepias syriaca

PODS (mid-Sept. into winter)
Pods are large with warty exteriors and erupt with feathery parachutes of seeds leaving soft, pale orange tongues clinging to silky yellow interiors. For darker stems and grayer exteriors, Milkweed may be picked well into winter.

• These pods have so much character. They are equally attractive used with stems in arrangements or with no stems in wreaths and craft projects.

Common Milkweed, Velvet-leaf, dried artichoke, Hickory leaves, on shredded cornhusk wreath

MULTIFLORA ROSE
Rosa multiflora

FLOWERS

Bloom: May–June To 6-ft. bush
Flowers: White, abundant in pyramidal
        clusters
Leaves: 7–9 leaflets on arching, trailing
        branches
Found: Roadsides, clearings, fencerows

PODS (late Sept.–Oct.)
As the bush slowly turns color, it blushes
with clusters of shiny crimson rose hips.
The later this pod is picked, the deeper is
the color, and the berries will also have
hardened.
• This glossy pod adds high color and a
jewel-like quality to Williamsburg and
mixed arrangements.

Multiflora Rose hips, Silver
Dollars

FLOWERS

Bloom: June–Sept. 1–3 ft. tall
Flowers: Pale lavender with yellow beaks
Leaves: Rough, widely toothed, prickly
stems
Found: Disturbed areas, sandy soil,
roadsides

HORSE-NETTLE
Solanum carolinense

PODS (late Sept.–Oct.)

Early fruits are yellow-green, but it is better to pick when the stems and leaves turn brown and berries are chrome yellow. Spraying minimizes the chances of rotting. It is a difficult pod to preserve but worth trying. Snip off any bad fruit.

• Even shriveled dry it is an interesting pod for the vivid yellow it adds to arrangements.

Horse-nettle, Panic grass

123

PALE SMARTWEED
Polygonum lapathifolium

FLOWERS

Bloom: July–Oct. 1–6 ft. tall

Flowers: Pink, sometimes greenish white, slender, dense arching clusters

Leaves: Narrow, lanceolate, on knotted stems

Found: Disturbed areas, fields, thickets, wet areas; common

PODS (late Sept.–Oct.)

Pennsylvania Smartweed, *P. pensyl-vanicum,* is another species similar to and equally familiar as Pale Smartweed. Its flower heads do not droop, however, and the stems are ruddy. Smartweeds are best picked when the entire plant turns to shades of auburn red. Do not remove the leaves when drying.

• The curl of the leaves on arching stems gives lovely action to these pods and adds sparkle to arrangements.

Pale Smartweed (Weber)

124

FLOWERS

Bloom: Aug.–Sept. 2–4 ft. tall
Flowers: Yellow-white in dense clusters toward top of stems
Leaves: Cloverlike on fine hairy stalks
Found: Open woods, sandy soil, prairies

ROUND-HEADED BUSH-CLOVER
Lespedeza capitata

PODS (late Sept.–Oct.)

Round-headed Bush-clover looks almost burlike from a distance. It can be picked from a green-brown stage to honey-brown tones. Picked late, pods become dusty gray-brown yet maintain their lovely bushy shape.

• In arrangements this pod adds a soft yet solid touch, and the tall stems work well in vertical florals.

Round-headed Bush-clover, squash seed flower

125

GUMWEED
Grindelia squarrosa

FLOWERS

Bloom: Aug.–Sept. 1–4 ft. tall
Flowers: Yellow, yellow disks, gummy
resin on bracts
Leaves: Linear, alternate, toothed
Found: Disturbed areas, railroad banks,
roadsides

PODS (late Sept. on)

Wait for this pod to lose all seeds and to turn a warm, sandy tan. It is worth rechecking to find the round, empty, shallow-cup heads with beautiful satiny bottoms. Usually this pod lightens with age in the field.

• Gumweed is an excellent, crisp, clean, honey-colored pod for any arrangement.

Gumweed, Common Plantain, Sweet Everlasting

FLOWERS

Bloom: Aug.–Sept. Up to 9 ft. tall
Flowers: Greenish white, upright, on arched branches
Leaves: Heart-shaped, alternate
Found: Disturbed areas, roadsides, wet places as well as dry

JAPANESE KNOTWEED
Polygonum cuspidatum

PODS (late Sept.–Oct.)

This bushy plant usually grows in clumps up to 9 ft. tall. Stems have a bamboolike appearance, growing straight and arching near the top. When the leaves drop, the entire bush becomes a striking cinnamon color and is lovely contrasted with winter grays.

• To best display the zigzag pods perched on arched branches, use sparingly; and for color use alone in a huge mass.

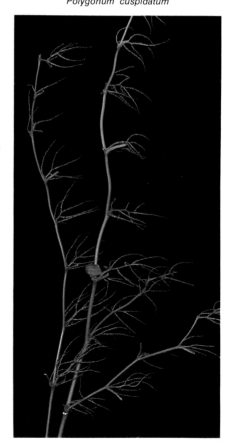

Japanese Knotweed, Queen Anne's Lace (Weiler)

FLOWERS

Bloom: Early Sept.–Oct. 1–7 ft. tall

Flowers: Deep pinks to dark purples, in large clusters

Leaves: Lanceolate, crowded, clasping hairy stems

Found: Fields, wet spots, thickets

NEW ENGLAND ASTER
Aster novae-angliae

PODS (end of Sept.–Oct.)

New England Asters are better picked in bloom and hung to dry. They burst into fuzzy seed heads and some retain the color of the bloom. Pods picked in the field are coarser-looking and are a darker brown.

● Take advantage of the soft muted colors; break the large clusters into smaller heads for all size arrangements.

New England Aster, Eucalyptus (Weiler)

FLOWERS

Bloom: July–Sept. 1–5 ft. tall
Flowers: Golden yellow, plumelike clusters
Leaves: Lance-shaped, sharp-toothed, crowded
Found: Meadows, clearings, roadsides; common

CANADA GOLDENROD-GALLS
Solidago canadensis

PODS (early Oct. on)

FOR INSECT GALLS: These unusual shapes, which look like onions on skewers, are caused by insects laying eggs. They are most attractive podlike forms and great in arrangements. The sleek, round galls in rich mahogany browns eventually drop the flower seed tops, and clean stems and galls remain.

• They are an excellent shape for contemporary containers and pottery.

Canada Goldenrod insect galls

FLOWERS
Bloom: Apr.–July 1–2 ft. tall
Flowers: Soft red spurs, yellow stamens, drooping, bell-like
Leaves: Deeply lobed, 3 separate leaves on each leaf stalk
Found: Shaded areas, rocky areas, dry woods

WILD COLUMBINE
Aquilegia canadensis

PODS (June–mid-Aug.)
Columbine continues to bloom a long time, so patience is needed to wait for the entire plant to ripen fully and turn brown. It has many branches with a lovely elongated, open tulip pod at the tip of each branch.

• Because this pod is nicely shaped and has a dainty quality, it can be used gracefully alone or in mixed bouquets.

Wild Columbine

FLOWERS

Bloom: May–June 1½–3 ft. tall
Flowers: Dull white, very small, open clusters
Leaves: Fernlike, deeply cut and toothed
Found: Woods, moist slopes, shade

SWEET CICELY
Osmorhiza claytoni

PODS (late June–Aug.)

The many Y-shaped, slender branches have long, black-hooked, split pods hanging at the tips. To avoid shedding, pick as soon as leaves wither and pods are black. Handle carefully, as pods catch on each other.

● There is a dainty, angular feeling about the shape of this plant; if possible, consider it as the center of interest in arrangements.

Sweet Cicely

131

FLOWERS

Bloom: Apr.–June 6–14 in. tall
Flowers: Yellow or crimson, tubular and
hooded, in broad spirals
Leaves: Long, soft, hairy, deeply incised
Found: Open sandy woods, thickets,
clearings

WOOD-BETONY
Pedicularis canadensis

PODS (July–Aug.)

Wood-betony (also known as Lousewort)
grows low and may be recognized by its
rosette of ferny foliage. Pods are dark with
lighter curled-up lips and are uniquely
sturdy-looking for such a small species.
Wait to collect until mature and brown in
the field.

• These almost woodlike, textured pods
are unusually perky and crisp for small
bouquets.

Wood-betony, Cinquefoil, acorns, pine
needles, Pearly Everlasting, pine cones,
Foxtail grass (Weber)

FLOWERS

Bloom: May–July 8–20 in. tall
Flowers: Greenish white, round umbels
on separate stalks, lower than
leaves
Leaves: 5 ovate, finely toothed leaflets to
each stem
Found: Dry, open, and moist woodlands

WILD SARSAPARILLA
Aralia nudicaulis

PODS (July–Aug.)

Early in the summer begin to look for
these delightful little berry balls, usually 3
to a stem and of varying heights. Some
may be picked green and hung to dry;
these dry to a greenish tan with little
ridges in the pods. Later the blue-berry
balls can be dried for dark accent.
• Even the bristly berryless ball of empty
stems contributes a fascinating pattern to
small arrangements.

Wild Sarsaparilla, Dried Solomon's-seal
leaves

FLOWERS
Bloom: May–Aug. 1–3 ft. tall
Flowers: White, or pink to purple, phlox-
like, 4 petals, clustered, fragrant
Leaves: Lanceolate, toothed, alternate
Found: Roadsides, edges of woods,
thickets

DAME'S ROCKET
Hesperis matronalis

PODS (July–Aug.)
As the flowers wither, 3- to 4-in.-long seed
pods of the Mustard family form up the
stem. Note the rhythmic, graceful quality
and delicate shape of the pods. Handle
with care.
• For a light touch include this airy fragile
pod in bouquets; it is also elegant used as
a single pod.

Dame's Rocket

FLOWERS

Bloom: May–July 2–6 ft. tall
Flowers: Shiny green outside, brown inside, small
Leaves: Lance-shaped, opposite, on ribbed stems
Found: Edges of woods, thickets, fencerows

FIGWORT
Scrophularia lanceolata

PODS (July–Aug.)

Figwort is a beautiful pod to observe. On short, fine branches off the main stem, smooth, nut-brown pods spiral-climb upward in polka-dot fashion. Figwort keeps well, is easy to handle, and is a good species to collect.

• This worthy pod may be used as a single stem in a special container; it also works beautifully in Williamsburg arrangements for background height.

Figwort, Narrow-leaved Cat-tail, Tansy, Tall Wormwood, garden Yarrow (Weiler)

135

ALUMROOT
Heuchera richardsonii

FLOWERS
 Bloom: Apr.–June 1–3 ft. tall
 Flowers: Greenish tan, slightly droopy
 Leaves: Rounded maple-leaf shape, low
 to ground
 Found: Dry woods, prairies

PODS (mid-July into Aug.)
Pods cling to tall, slender stems in sets of 2 or 3 upturned bells. Pick as soon as stems turn brown, early for lighter color and later for darker browns.

• Alumroot is an attractive, dainty pod and may be used with long stems or cut short for use in small arrangements, as illustrated in the adjacent photo.

Alumroot, Strawflowers (Weiler)

FLOWERS

Bloom: Apr.–June 8–20 in. tall
Flowers: Light pink or white, petals swept
back, stamens form pointed
beaks
Leaves: Blunt, lance-shaped, basal,
leaves sometimes reddish
Found: Open woods, meadows, moist
slopes, prairies

SHOOTING STAR
Dodecatheon meadia

PODS (mid-July–Aug.)

Shooting Star pods are special! Shooting
out from the top of a single main stem are
handsome, elongated, upward-reaching
cups with lighter sawtooth tips. Allow
these pods to dry in the field until they are
deep mahogany brown.

• This pod is such a pleasing shape. Use it
sparingly in arrangements to accentuate
its beauty.

Shooting Star

FLOWERS

Bloom: May–Aug. 1–2 ft. tall
Flowers: White, 5 petals, pale green
 burlike centers
Leaves: Usually in 3–5 leaflets on slightly
 zigzag stems
Found: Edges of woods, medium-dry
 woods, thickets, meadows

WHITE AVENS
Geum canadense

PODS (mid-July–Oct.)

The seeds fall early, leaving small, blunt
heads at the end of very slender, slightly
drooping branches. The angular thrusts of
the branches give these pods a soaring
quality. Bunch carefully, as they tend to
stick together.

• For linear beauty use just one stem of
Avens, or make an unmixed arrangement
for a dramatic effect.

White Avens

FLOWERS

Bloom: May–June 8–20 in. tall
Flowers: Lemon yellow, bell-shaped,
         droop straight down
Leaves: Pointed oval, stems appear to
         pierce leaves
Found: Moist woods, thickets

PERFOLIATE BELLWORT
Uvularia perfoliata

PODS (mid-July on)

Perfoliate Bellwort is ready to pick and
easily noticed in the woods when the
graceful leaves have dried parchment tan,
pierced by small, solid, lily-shaped pods.

• Consider the entire plant for arrangements, as the flowing, rhythmic leaves are
as beautiful as the pod. The small pods
may also be used separately in miniatures.

Perfoliate Bellwort

139

YELLOW AVENS
Geum aleppicum

FLOWERS
Bloom: June–Aug. 2–4 ft. tall
Flowers: Yellow, petals shorter than sepals
Leaves: Large leaflets on slightly hairy stems
Found: Low meadows, thickets, shaded areas

PODS (Aug.–Sept.)
These pods have characteristics similar to those of the White Avens. However, they have much larger burlike heads on slightly hairy, fragile stems.

● The beauty of this species is the graceful branching, and pods should be arranged to focus on the rhythmic lines.

Yellow Avens

FLOWERS

Bloom: July–Aug. 1–3 ft. tall
Flowers: Lavender, in pairs up stems
Leaves: Oval, opposite, corasely toothed
Found: Woods, thickets, partial shade

LOPSEED
Phryma leptostachya

PODS (Aug.–Sept.)

The identifying characteristic of Lopseed is the unusual way the pods lop downward against the stem in pairs, giving it a winged pattern. Pods are almost black in color, and usually there are 2–5 spikes branching from the main stem.

• The spikes are so slender and trim they resemble black sticks, as they dart out of bouquets, producing a staccatolike accent.

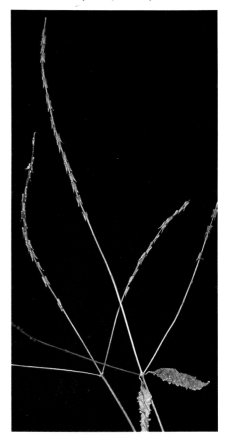

Lopseed, St. Johnswort, feathers

141

NEW JERSEY TEA
Ceanothus americanus

FLOWERS
Bloom: June–Aug. 1–3-ft. shrub
Flowers: Creamy white, small, feathery clusters
Leaves: Ovate, pointed, alternate on bronze stems
Found: Dry woods, sandy fields, thickets

PODS (late Aug. on)
The first noticeable pods are small, green, and 3-lobed; later they turn blue-black, finally drying to brown. Pick some seedpods for the "grape-bunch" look; later in the season pods drop off and mini shallow saucer shells remain. These crisp, clustered, dark-ringed saucers atop dark, branching stems are charming.
● Both attractive shapes are worth collecting for creative arrangements.

New Jersey Tea, Bur-reed, garden Yarrow

FLOWERS

Bloom: June–Sept. Climbing vine
Flowers: Greenish white, clustered in leaf axils
Leaves: Deeply cut, like very pointed maple leaves
Found: Rich soil, riverbanks, thickets

WILD CUCUMBER
Echinocystis lobata

PODS (late Aug. into Sept.)

Vines are laden with parchment-colored, tulip-shaped pods covered with soft prickles. Note the interesting deep seed pockets in the pods.

• For arrangements, the vines are too weak to hold pods, but heads can be attached to false stems. For a different approach, fill an unusual bowl with this thorny fruit.

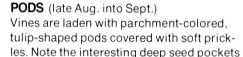

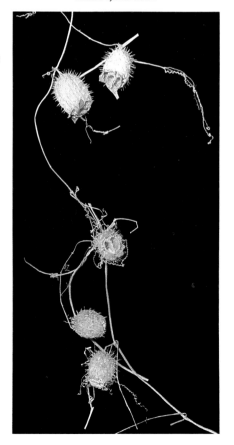

Wild Cucumber, Yellow Rocket, dried artichoke (Weber)

143

SOLOMON'S-SEAL
Polygonatum biflorum

FLOWERS

Bloom: May–June 1–3 ft. tall
Flowers: Greenish yellow, bell-shaped, in pairs, dangle from leaf axils
Leaves: Ovate, alternate, smooth
Found: Woods, thickets, moist sandy soil

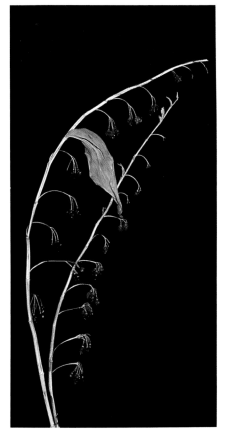

PODS (Sept. on)

Wait for the leaves to become pale and dry and for berries to be blue-black. Many times the stems are leafless by the time the pods are found. Pick and set upright in a container, allowing berries to droop while drying.

• The hanging fruit on the graceful curve of the stem provides a different shape and color for arrangements.

Solomon's-seal, Carrion-flower

FLOWERS

Bloom: May–July 2–3 ft. tall
Flowers: Creamy white, loose pointed
 clusters at top of stems
Leaves: Oval-lanceolate, alternate
Found: Woods, thickets, clearings

FALSE SOLOMON'S-SEAL
Smilacina racemosa

PODS (Sept.–Oct.)

When ruby-red berries appear, pick and hang upside down to dry and to harden stems. Ideally, leaves by then are not too mottled. Handle carefully and consider spraying after drying to avoid losing berries.

• Like Solomon's-seal, the leaves sometimes develop beautiful curled shapes and pale colors to include in bouquets.

False Solomon's-seal, dyed imported Centaurea, sprayed Sea Lavender

145

CULVER'S-ROOT
Veronicastrum virginicum

FLOWERS
Bloom: June–Sept. 2–6 ft. tall
Flowers: White, tubelike, slender racemes
Leaves: Slender, sharp-toothed, in
　　　　whorls of 3–7, encircle stems
Found: Thickets, wet meadows, prairies,
　　　　woods

PODS (Sept.–Oct.)
Culver's-root stands out noticeably in the field with its elongated, very slender spike clusters; however, it can be easily mistaken for Blue Vervain. Note that Culver's-root does not have as many spikes as Blue Vervain and pod heads branch in whorls.
● Spikes picked green are denser than the dark matured pods, but both shapes are striking in arrangements.

Culver's-root, miscellaneous
grasses

FLOWERS

Bloom: Aug.–Oct. 2–5 ft. tall
Flowers: White to pale blue-lavender rays, in pointed clusters
Leaves: Arrow-shaped, toothed, narrow
Found: Open woods, shaded roadsides, clearings

ARROW-LEAVED ASTER
Aster sagittifolius

PODS (Sept.–Nov.)

There are over 50 species of wild Asters in the northeastern United States (only 4 mentioned in this book). They bloom in various colors, but all, characteristically, have fuzzy heads and similar strawflowers remain when seeds fall; the silhouettes, however, differ. Arrow-leaved Asters have off-white seed heads in open, conical shapes.

● All Asters adapt well in mixed arrangements.

Arrow-leaved Aster, Black-eyed Susan (Weiler)

147

POKEWEED
Phytolacca americana

FLOWERS
Bloom: July–Sept. 4–10 ft. bush
Flowers: Greenish white, on slender racemes
Leaves: Lanceolate, large, alternate
Found: Open woods, damp thickets, roadsides, clearings

PODS (early Sept. on)
These slender, drooping clusters of berry pods offer a variety of picking choices; gathered early as they begin to pod, the color graduates from purple-pink to pale pea-green tips. Picked later and dried, heavy, ripe, purple-black berries are retained; when all berries are gone, vivid raspberry-colored skeletons are available.
• Collect all forms for special and varying effects in arrangements.

Pokeweed, Summer Cypress, pink Statice (Weber)

FLOWERS

Bloom: June–Sept. 3–5 ft. tall
Flowers: White, tiny, sparse blooms, on long, very slender spikes
Leaves: Lanceolate, in pairs, toothed
Found: Rich thickets, edges of woods

WHITE VERVAIN
Verbena urticifolia

PODS (Sept.–Nov.)

This pod is much more exciting when picked green and dried. The naturally dried specimens turn gray-brown and do not have the sparkle of the green pods.

• These very slender, very long, finely podded spikes that seem to be almost animated are striking used alone in a huge mass. Pods may be shortened and applied in bunch accents to small arrangements.

White Vervain, dried Smooth Dock, Horsetail (Weber)

FLOWERS
Bloom: July–Oct. 1–3 ft. tall
Flowers: White, fuzzy, in open clusters
Leaves: Heart-shaped, opposite
Found: Woodlands, thickets, rich soil

WHITE SNAKEROOT
Eupatorium rugosum

PODS (late Sept.–Oct.)
It is better to pick White Snakeroot as a flower and hang to dry. The flowers burst into fuzzy, gray-white pod heads; if they are allowed to dry in the woods, the pods are apt to look nondescript. Spray before arranging.
• The feathery, light heads give a lacy, loose touch in dainty containers.

White Snakeroot, Round-headed Bush-clover (Weber)

FLOWERS

Bloom: July–Aug. 2–3 ft. tall
Flowers: Greenish white, 5 petals (sepals) on individual stems
Leaves: Palmate, 3 leaflets, deeply toothed
Found: Dry open woods, wooded roadsides, thickets

THIMBLEWEED
Anemone virginiana

PODS (mid-Sept. on)

Thimbleweed pods stay green a long time; some should be gathered for that particular lovely pale color. When the pods attain a creamy color, pick and spray to prevent them from bursting into mangy cotton heads. Some may show just a tuft of cotton at the tip, which adds interest to the textural heads.

• This charming "thimble" is most appropriate for colonial bouquets and dainty miniatures.

Thimbleweed, sedges

TALL UPLAND BONESET
Eupatorium altissimum

FLOWERS
Bloom: July–Sept. 2–5 ft. tall
Flowers: Off-white, fuzzy-tipped, in clusters
Leaves: Slender with 3 main veins, often smaller leaves in axils
Found: Woodlands, northern areas, uplands

PODS (mid-Sept.–Oct.)
Tall Upland Boneset picked in bloom bursts into beautiful, pale green, fuzzy seed heads as it dries. For shiny, pale lime-green faces, shake the seedpods. When it matures in the field, it turns a soft beige color. This is generally true of about 26 species of thoroughworts (*Eupatorium*).
• The pastel colors of Boneset are welcome additions to colonial or Williamsburg arrangements.

Tall Upland Boneset (Weiler)

FLOWERS

Bloom: July–Sept. 1–5 ft. tall
Flowers: Yellow, tubed bell-shaped
Leaves: Smooth, short leaf stalks,
opposite
Found: Open dry woods, edges of woods

SMOOTH FALSE FOXGLOVE
Gerardia laevigata

PODS (mid-Sept.–Oct.)

These pods stay green a long time and finally turn black on soft gray-brown stems. Pods are fat with dainty, curved beaks which split open; they are beautiful pods deserving close scrutiny of their design.

• This graceful seed container gives an unusually fine black accent to the usual browns of other dried wildflowers. Used alone it is also very attractive.

Smooth False Foxglove,
False Boneset

153

NARROW-LEAVED CAT-TAIL
Typha angustifolia

FLOWERS
Bloom: May–July 2–5 ft. tall
Flowers: Yellow male spikes separated by short gaps from green female spikes
Leaves: Long, bladelike, erect
Found: In dense stands in marshes, shallow water; common

PODS (mid-July–Aug.)
Pick as soon as pollen bloom disappears and clean tip remains. Common Cat-tail, *Typha latifolia,* is too large and overwhelming for most arrangements.

• This slender, smaller species is very popular and is used a great deal in fall bouquets. *Be sure to spray Cat-tails,* as they burst open indoors and are then a total loss as a usable pod.

Narrow-leaved Cat-tail, Wild Yarrow

FLOWERS

Bloom: May–Aug. 1–4 ft. tall
Flowers: Greenish brown, burlike round
         heads
Leaves: Irislike, linear, erect
Found: Muddy shores, shallow water,
       edges of rivers; common

BUR-REED
Sparganium americanum

PODS (July–Aug.)

These pods may be picked all summer for colors ranging from sulfur greens to amber browns and for the attractive clusters of seeds with beaks that form studded-ball pods. Hang to dry and to harden the stems.

• Be sure to utilize the leaves, as they add graceful linear movement to arrangements.

Bur-reed, Narrow-leaved Cat-tail

MEADOWSWEET
Spiraea latifolia

FLOWERS
Bloom: June–Sept. 2–5-ft. shrub
Flowers: Pale pink or white, loose pyramid clusters
Leaves: Oval, alternate, coarsely toothed on reddish stems
Found: Roadsides, lowlands, damp meadows; especially northern areas

PODS (mid-Aug.–Sept.)
These feathery heads are noticeable on bushy shrubs throughout the summer. In fall the light-brown-sugar color and airy, tapered heads make an interesting addition to the pod collection.
• These pods can be used when spikes are not desired and yet some contrast from round and ball shapes is needed in arrangements.

Meadowsweet, potpourri of pods (Weber)

FLOWERS

Bloom: June–Aug. 4–8 ft. tall
Flowers: White, crowded, in flat-topped umbels
Leaves: Palmate, huge, often over 1 ft. in size, ridged hollow stems
Found: Moist areas, disturbed areas, low meadows

COW-PARSNIP
Heracleum maximum

PODS (Aug.–Sept.)

This is a big plant. The umbel pod head may measure up to 12 in. in diameter. Look early to find it with seeds, as the textural beauty is in the watermelon-size seedpods with dark lines radiating from a notch.

● If you are thinking big, think of Cow-parsnip. It makes impressive commercial and display arrangements or is attractive as one large pod.

Cow-parsnip

157

PURPLE LOOSESTRIFE
Lythrum salicaria

FLOWERS
Bloom: June–Sept. 2–6 ft. tall
Flowers: Magenta-purple, crowded on spikes
Leaves: Lanceolate, stalkless, opposite
Found: Swamps, wet meadows, river areas; in great patches

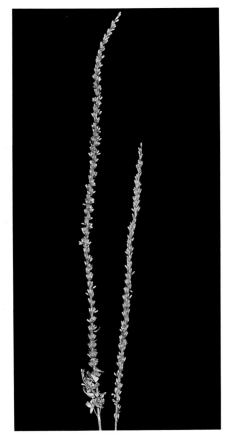

PODS (Aug.–Sept.)
Purple Loosestrife is easily identified by the magenta spikes that carpet large expanses of wet areas. If they are picked early in the podding stage, a rosy tinge may be captured. Later the upright seeds in spicy brown tones thickly encircle the spikes in whorls.

• There is a popcorn-kernel quality to the pod spike that adds interesting texture and height to arrangements.

Purple Loosestrife, dried Common Milkweed leaves, Japanese Lanterns (Weiler)

FLOWERS

Bloom: July–Sept. 2–5 ft. tall
Flowers: White, loose umbel heads
Leaves: Long leaflets, slender, usually
        not toothed
Found: Wet areas, swamps, moist woods

COWBANE
Oxypolis rigidior

PODS (Aug.–Sept.)

Cowbane has pod clusters at the ends of umbrellalike ribs that burst out from the top of the main stem like an exploding fireworks display. There is an exciting flair to this pod enhanced by the burnished pecan color of the very smooth stems.

• The ornamental quality of Cowbane lies in its clean, simple, soaring lines.

Cowbane, metal flowers

FLOWERS
Bloom: July–Sept. 1–5 ft. tall
Flowers: Purple-blue, 5 petals, branching, pencil-size spikes
Leaves: Lanceolate, opposite, coarsely toothed on grooved stems
Found: Low roadsides, moist thickets, shores; common

BLUE VERVAIN
Verbena hastata

PODS (mid-Aug.–Oct.)
The neat, tight, almost braided look of the Blue Vervain pod is easy to spot by its multiple pencil-like spikes. It matures to deep brown and has a nicely structured stem. Hoary Vervain, *V. stricta,* is a larger species with purple flowers and a much coarser texture.

• These clustered, slender spikes will emphasize the vertical lines in any arrangement.

Blue Vervain (Weiler)

FLOWERS

Bloom: July–Sept. 4–8 ft. tall
Flowers: White, round-headed, in umbel clusters
Leaves: Lower leaves very large, compound leaflets on smooth purple stems
Found: Wetlands, edges of woods, along streams

ANGELICA
Angelica atropurpurea

PODS (mid-Aug.–Sept.)

Pick this lovely, nutmeg-colored pod early because seeds drop quickly and the plants become scraggly. Spray to keep seeds on umbrellalike heads.

• Combined with Cat-tails and grasses, these large ornamental pods make attractive floor or corner arrangements.

Angelica, Common Milkweed, Narrow-leaved Cat-tail

WILLOW-HERB
Epilobium coloratum

FLOWERS
Bloom: July–Oct. 1–3 ft. tall
Flowers: Pink or white, very tiny,
numerous, often nodding
Leaves: Lanceolate, toothed, sometimes
ruddy stems
Found: Wet meadows, lowlands

PODS (late Aug.–Sept.)
These sprightly little flowers become very numerous, elongated, stiff pods (like Fireweed pods) that burst open, filled with silky white seeds that float away. Many of these plants are so thick with pod skeletons that they look like feathery tumbleweed bushes.

• Willow-herb broken into smaller branches fills the arrangements with wispy touches.

Willow-herb, sliced dried Osage Orange, dried Willow leaves

FLOWERS

Bloom: July–Sept. 1–3 ft. tall
Flowers: White, often pink-tinged at tips, turtle-head-shaped
Leaves: Narrow, toothed, in pairs
Found: Low ground, stream banks, swamps

TURTLEHEAD
Chelone glabra

PODS (late Aug.–Oct.)

Turtlehead, as the name implies, has flowers and then also pods that look like turtle heads. In a dense cluster these charming shapes hug the stem near the top. The crisp shapes and variations of brown tones give a wonderful three-dimensional quality to these pods.

• There is a stateliness about Turtlehead that invites special attention in arrangements.

Turtlehead, Plume grass

163

FLOWERS

Bloom: May–June 1–3 ft. tall
Flowers: Violet-purple, down-curved
    sepals veined in white or yellow
Leaves: Sword-shaped, grasslike, erect
Found: Marshes, edges of rivers,
    wetlands

WILD IRIS
Iris virginica

PODS (Sept.–Oct.)

Keep careful watch for pods to turn brown
and to drop their seeds, leaving an open,
lily-shaped pod. They often remain closed
until frost, and colors are sometimes drab;
however, the shape is worth highlighting
with pecan stain.

• The unusual nutlike quality of this pod
enhances mixed bouquets, or emphasize
its beauty with a few used alone in a bud
vase.

Wild Iris, Butter-and-eggs, Statice,
Baby's Breath (Weiler)

FLOWERS

Bloom: June–Sept. 1–3 ft. tall
Flowers: White, tiny, in whorls, on loosely branched stalks
Leaves: Ovate, basal, upright, shiny green
Found: Shallow water, swamps, muddy shores

WATER-PLANTAIN
Alisma triviale

PODS (Sept.–Oct.)

Pick these pods when brown. You may need boots to harvest them, as these worthwhile pods prefer wet soil. These exquisite, amber-brown, miniature pods on hair-fine stems almost bounce in a dotted pattern.

• Used alone in a special china container, Water-Plantain makes a charming, airy bouquet; or it can be used in place of Baby's Breath in colonial arrangements.

Water-Plantain

FLOWERS

Bloom: July–Aug. 2–4 ft. tall
Flowers: Deep rosy pink, numerous, in small umbels
Leaves: Long, lanceolate, opposite, smooth
Found: Wet meadows, swampy areas

SWAMP MILKWEED
Asclepias incarnata

PODS (early Sept.–Oct.)

Swamp Milkweed differs from Common Milkweed in its branching characteristic plus its lighter, more aesthetic pod. It is smooth-stemmed and sandy-colored and has the butterfly pods of the Whorled Milkweed.

● These graceful, artistic branches with pods poised in terminal clusters are well suited to contemporary or asymmetrical arrangements.

Swamp Milkweed

FLOWERS

Bloom: July–Sept. 2–7 ft. tall
Flowers: Pinkish red, flat-topped clusters
Leaves: Lanceolate, toothed, 4–5 in
whorls along ruddy spotted
stems
Found: Damp areas, wet thickets, shores

SPOTTED JOE-PYE-WEED
Eupatorium maculatum

PODS (mid-Sept.–Oct.)

For lovely red-brown, fuzzy heads, pick the flowers and hang to dry. Like other *Eupatoriums,* they develop richer colors when dried indoors; matured in the field, they become ordinary beige-brown in color.

• Sweet Joe-Pye-weed, *E. purpureum* (pink), and Hollow Joe-Pye-weed, *E. fistulosum* (more domed), have similar pod qualities, and all specimens have handsome, workable pod heads.

Joe-Pye-weed, dyed imported Centaurea, Ruscus leaves (Weiler)

167

FLOWERS
Bloom: July–Sept. 1–1½ ft. tall
Flowers: White, 3 round petals in whorls of 3 along stems
Leaves: Arrow-shaped, large, basal
Found: Shallow water, muddy areas, edges of rivers; common

BROAD-LEAVED ARROWHEAD
Sagittaria latifolia

PODS (early Sept.–Oct.)
The stems are soft and watery, which makes them tend to fall over as they mature. It is better to pick the brown seedpods while the stems are still green, and hang to dry and harden. The graduated podballs have an interesting linear texture.
• In arrangements it may be necessary to support the stems with wire.

Broad-leaved Arrowhead, Sweet Everlasting, Sea Lavender (Weiler)

168

FLOWERS

Bloom: July–Sept. 1–2 ft. tall
Flowers: White, small, clustered in leaf axils
Leaves: Deeply cut, in pairs horizontally
Found: Wet soil, low meadows, near water

WATER-HOREHOUND
Lycopus americanus

PODS (end of Sept.–Oct.)

This plant is impressive because of its lovely linear quality with little, dark, burlike pods threaded on a square stem. The pods, like beaded jewels, are graduated in size to a fine point.

• For that aesthetic touch, use an unusual container and carefully arrange these flowing, rhythmic pods.

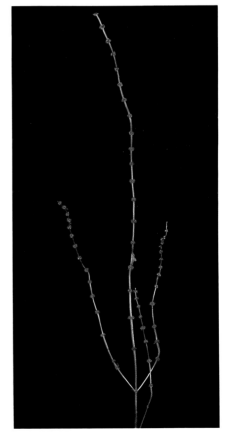

Water-horehound, Catalpa pods

PURPLE GERARDIA
Gerardia purpurea

FLOWERS
 Bloom: Aug.–Oct. 6–30 in. tall
Flowers: Pink, inflated, bell-like, in leaf
axils
 Leaves: Linear, fine, in pairs
 Found: Damp sandy soil, shores

PODS (early Oct.–Nov.)
This is a late-maturing pod that changes
into lovely fall colors, from deep red-
browns in early fall to rich warm browns.
The tiny, smooth rosebud pods appear *in
pairs* tucked in the leaf axils of the slender
stems. Let this pod mature before picking.
• This is a delightful pod for delicate
bouquets and is especially desirable for
small containers.

Purple Gerardia (Weber)

GRASSES, SEDGES AND RUSHES
AND WINTER SKELETONS

GRASSES, SEDGES AND RUSHES

Grasses usually are found growing in clumps with a variety of colors ranging from bleached beige, wheat golds, pale greens, and warm browns, to grayed purples, adding unexpected swaying color patches to the subdued fall and winter landscape.

The variety of shapes adds a lovely wispy accent and movement to dried arrangements, along with the curled, bleached leaves of some species. Hang grasses upside down in bunches to dry in order to prevent the heads from drooping.

Sedges are distinguished from grasses by having 3-sided solid stems and no stem joints. They usually grow in clumps or masses, in marshes or in swamp areas, but some species may also be found on roadsides and prairies.

Rushes have round, usually pulpy stems and no joints. They too like the wet habitat of sedges. Some species to look for:

1. MILLET (*Panicum*)
2. TIMOTHY GRASS (*Phleum*)
3. RED TOP (*Agrostis*)
4. REED CANARY GRASS (*Phalaris*)
5. BROMESEDGE (*Andropogon*)
6. SWITCH GRASS (*Panicum*)
7. LITTLE BLUE STEM (*Andropogon*)
8. BIG BLUE STEM (*Andropogon*)
9. SUDAN GRASS (*Sorghum*)
10. REED (*Phragmites*)
11. FOXTAIL (*Setaria*)
12. COMMON BROME GRASS (*Inermis*)
13. GREEN BULRUSH (*Scirpus*)
14. HARD ROUNDSTEM BULRUSH (*Scirpus*)
15. NUTGRASS (*Cyperus*)
16. BOTTLEBRUSH SEDGE (*Carex*)

2   3   4

5   6   7   8

10

11

12

13 14 15 16

WINTER SKELETONS

Many weed pods remain beautiful in spite of winter winds and weather. They lose their warm earth tones and acquire a patina of dark gray to silver white. Many change their shapes as winds whip away parts of their structure, leaving a lovely wispy skeleton. Some species to look for:

1. QUEEN ANNE'S LACE has fine stark spikes radiating from the center; it resembles a fragile snowflake.
2. MILKWEED pods turn silvery gray on black stems.
3. The insect gall on CANADA GOLDENROD looks like a silver onion skewered onto the stem.
4. MULLEIN pods darken to a black-brown with rosette tips frosted white.
5. EVENING-PRIMROSE pods turn deep brown on even darker brown stems. The split-open pods reveal silver-white interiors.
6. YELLOW ROCKET develops smooth gray stems with protruding spines.
7. FROST ASTER loses its fuzzy seeds but retains a strawlike flower.
8. GOLDENROD'S fluff is blown away, leaving a silver gossamer skeleton.
9. CHICORY pods become blunted by the winds, but lovely, rhythmic branches remain.
10. YELLOW GOAT'S-BEARD's seed head flies away, leaving a blunt head on an unusual branched stem structure which is thick at the top tapering to a fine stem at the bottom.
11. WILD LETTUCE becomes gray with white pods polka-dotted up the stem.
12. CARRION-FLOWER loses all berries, and a porcupine ball remains.

The skeletons of winter add new color and patterns for collecting and arranging. There are many other forms you will discover for yourself. This list is but a suggestion to stimulate your interest in the winter season.

1 2 3 4

5   6   7   8

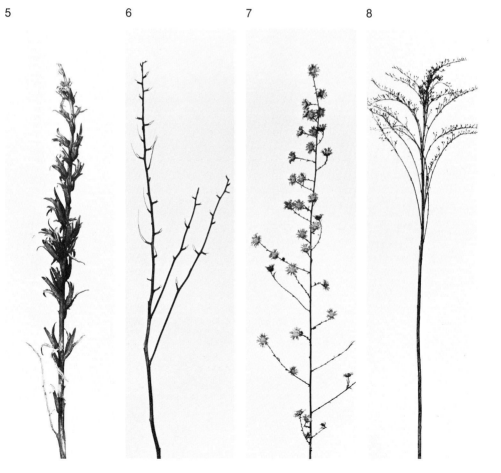

9   10   11   12

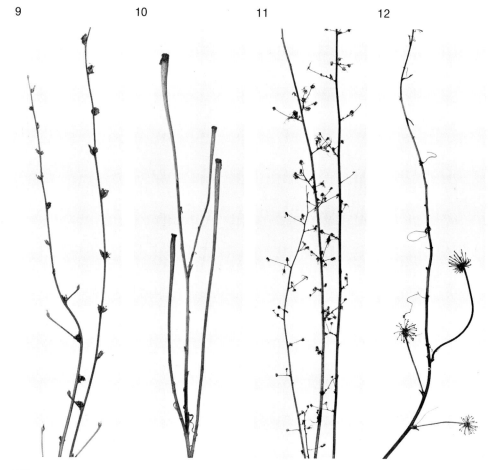

GLOSSARY

Alternate leaves Leaves not opposite each other (Garlic Mustard).
Axil The angle formed by the leaf and the stem; the junction of leaf and stem.
Basal leaves Leaves at base of stem (Water-plantain).
Botanical names The "universal language" for designating a particular kind of plant, consisting of the genus and species names.
Bracts Modified leaves, usually at the base of the flower (Gumweed).
Calyx The usually green outer circle of the flower, composed of leaflike sepals (Thimbleweed) or a saclike envelope (Bladder Campion).
Composite A close cluster of flowers that appear as one head or single blossom (Knapweed or Yarrow).
Compound leaf A leaf divided into separate smaller leaflets (Sweet Cicely).
Disturbed land Land that has been turned over, such as farmland or land for building and road purposes, or where trees have been removed, leaving open spaces.
Inflorescence The flower cluster.
Lanceolate leaves Lance-shaped, much longer than wide, tapering toward the tip (Lance-leaved Goldenrod).
Lobed leaves Margins deeply indented, sometimes almost to the midrib (Compass-plant or Wild Lettuce).
Node The point where leaves are attached to stem.
Opposite leaves Leaves in pairs, at the same level on opposite sides of stem (Catnip).
Ovate leaves Leaves somewhat egg-shaped, broader at base (Spreading Dogbane).
Palmate leaf Simple or compound, with lobes or leaflets radiating from a central point (Wild Cucumber, simple; Rough-fruited Cinquefoil, compound).
Petiole The leaf stalk.
Pinnate leaf Compound with leaflets in pairs or alternately arranged along

a center rib (Leadplant).

Pod The seed vessel or seed container of the plant.

Prairie The French word for meadow. A plant cover of grasses and wildflowers without trees or shrubs.

Raceme Flowers arranged along the stem, each flower with its own individual stalk (Fireweed).

Rosette A circular cluster of leaves at base of the plant (Shepherd's Purse).

Serrated leaf Margins of leaf are sharp-toothed, resembling a saw edge (Prairie Rose).

Sessile Leaf without a stalk (Solomon's-seal).

Species A distinct kind of plant.

Spike Flowers that are stalkless or nearly stalkless arranged along a stem (Common Plantain).

Toothed A leaf with small projections, either sharp or blunt, along margin (Wild Bergamot).

Umbel Flower stalks all radiating from the same point like ribs of an umbrella (Queen Anne's Lace).

Weeds Weeds are actually all the uncultivated plants that are scorned by man.

Whorl The leaves or flowers arranged in a circle around the stem, radiating from the same point (Spotted Joe-Pye-weed leaves, Flowering Spurge).

Wildflowers The flowers of the uncultivated plants.

INDEX

Wildflowers and weeds are listed here according to both their botanical names (in *italics*) and their most popular common names.